T0211375

BestMasters

Mit „BestMasters" zeichnet Springer die besten Masterarbeiten aus, die an renommierten Hochschulen in Deutschland, Österreich und der Schweiz entstanden sind. Die mit Höchstnote ausgezeichneten Arbeiten wurden durch Gutachter zur Veröffentlichung empfohlen und behandeln aktuelle Themen aus unterschiedlichen Fachgebieten der Naturwissenschaften, Psychologie, Technik und Wirtschaftswissenschaften. Die Reihe wendet sich an Praktiker und Wissenschaftler gleichermaßen und soll insbesondere auch Nachwuchswissenschaftlern Orientierung geben.

Springer awards "BestMasters" to the best master's theses which have been completed at renowned Universities in Germany, Austria, and Switzerland. The studies received highest marks and were recommended for publication by supervisors. They address current issues from various fields of research in natural sciences, psychology, technology, and economics. The series addresses practitioners as well as scientists and, in particular, offers guidance for early stage researchers.

Erik Kalz

Diffusion under the Effect of Lorentz Force

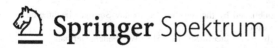 Springer Spektrum

Erik Kalz ⓘ
Finsterwalde, Germany

ISSN 2625-3577   ISSN 2625-3615   (electronic)
BestMasters
ISBN 978-3-658-39517-9   ISBN 978-3-658-39518-6   (eBook)
https://doi.org/10.1007/978-3-658-39518-6

Responsible Editor: Marija Kojic
This Springer Spektrum imprint is published by the registered company Springer Fachmedien
Wiesbaden GmbH, part of Springer Nature.
The registered company address is: Abraham-Lincoln-Str. 46, 65189 Wiesbaden, Germany

Abstract

English[1]

It is generally believed that collisions of particles reduce the self-diffusion coefficient. In this thesis, we show that in systems under the effect of Lorentz force, which are characterized by diffusion tensors with antisymmetric elements, collisions surprisingly can enhance self-diffusion. In these systems, due to an inherent curving effect, the motion of particles is facilitated, instead of hindered by collisions. Consistent with this we find that the collective diffusion remains unaffected. Using a geometric model, we theoretically predict a magnetic field governed crossover from a reduced to an enhanced self-diffusion. The physical interpretation is quantitatively supported by the force autocorrelation function, which turns negative with increasing the magnetic field. Using Brownian-dynamics simulations, we validate the predictions.

Deutsch

Man nimmt allgemein an, dass Teilchenkollisionen den Selbstdiffusionskoeffizienten verringern. In dieser Arbeit wird gezeigt, dass in Systemen unter dem Einfluss einer Lorentzkraft, welche durch antisymmetrische Nebendiagonalelemente im Diffusionstensor gekennzeichnet sind, Teilchenkollisionen die Selbstdiffusion erhöhen können. Die charakterischen Nebendiagonalelemente sorgen direkt dafür, dass Kollisionen zwischen Teilchen die Selbstdiffusion nicht

---

mehr notwendigerweise verringern. Konsistent mit dem hier vorgeschlagen physikalischen Mechanismus bleibt die kollektive Diffusion von der Lorentzkraft unbeinflusst. Mittels eines geometrischen Modells wird in dieser Arbeit ein Magnetfeld-getriebener Übergang von Reduktion der Sebstdiffusion durch Teilchenkollisionen hin zu Verstärkung durch Kollisionen hergeleitet. Das physiklaische Bild zur Erklärung des Effektes wird quantitaiv durch die Kraft-Autokorrelationsfunktion unterstützt, welche mit steigender Magnetfledstärke negativ wird. Die Vorhersagen der Theorie werden durch Brownsche-Dynamik Simulationen bestätigt.

Contents

List of Figures

Introduction

1

Diffusion as a phenomenon has had a great impact on the development of physics. In 1827 the botanist Robert Brown observed the phenomenon of moving pollens in a solvent [2, 3]. This discovery led to vehement scientific debates. The two predominant viewpoints on the question of the existence of atoms, both, in alternating fashion interpreted the ongoing experimental work on the so-called Brownian motion in their interests. 80 years of debate produced almost no scientific progress, as it was not clear which quantity to measure in experiments [4]. Only when Albert Einstein in 1905 [5] came up with relating the mean-squared displacement of the particle to a diffusion coefficient, he explained Brownian motion as a diffusive process. Shortly after Einstein, in 1908 Jean-Baptiste Perrin [6] experimentally probed Einsteins predictions and thus finished the long debate about the existence of atoms.

Taking a different route, in 1906 Marian von Smoluchowski [7] drew the same conclusion for the mean-squared displacement of the Brownian particles, but with mathematically more profound arguments than Einstein did shortly before him. Both of these theories later were pursued by Max Planck and Adriaan Fokker, who discovered the general probabilistic equation for Brownian movement [8].

The nowadays well-understood Fokker-Planck equation provides a probabilistic description of the particle performing random motion. One can find exact analytical solutions to the equation for the case of non-interacting particles. Though interactions can be included in a rather straightforward fashion, the resulting Fokker-Planck equation for the many-body systems is not suitable for theoretical analysis. Despite other very successful theories dealing with many-body systems, just to name (dynamical) density functional theory [9] as one, recently a new method has been proposed by Maria Bruna and Jonathan Chapman [10–12] to deal with particle interactions in the Fokker-Planck description. They were able to treat hard-core interactions, one of the most intensively studied inter-particle interaction models, via the geometry of excluded volume. One of the advantages of their dynamical

E. Kalz, *Diffusion under the Effect of Lorentz Force*, BestMasters, https://doi.org/10.1007/978-3-658-39518-6_1

theory is that they are neither restricted to the thermodynamic limit nor deviations close to equilibrium, as this was the case in earlier, related work [13].

The geometric method is well-suited for particles undergoing isotropic diffusion. In this case, the diffusion coefficient is a scalar. It is generally assumed that isotropic diffusion is solely described by scalar diffusion coefficients. However, there exist systems undergoing isotropic diffusion, which cannot correctly be described by a scalar coefficient. One exemplary system in soft matter is active chiral particles. But the system, on which this thesis focuses, is charged particles diffusing under the effect of Lorentz force.

An applied magnetic field causes a Lorentz force, which curves the trajectory of a charged particle. The effect of Lorentz force on a diffusing Brownian particle is of a slightly different nature. The motion of Brownian particles is modelled as an overdamped diffusive process. There the concept of velocity is not defined and instead, the affected quantity by the Lorentz force is the probability flux in the Fokker-Planck description. In 2018 Hyun-Myung Chun et al. [14] showed that the probability flux has an additional rotational component, which is induced by the magnetic field. For an illustration see Fig. 1.1. Furthermore, they showed that the usual procedures in deriving the probabilistic Fokker-Planck equation do not work, as they fail to reproduce these rotational probability fluxes.

Chun et al. essentially stated that the system of charged Brownian particles under Lorentz force has to be described by tensorial diffusion. The effect of the Lorentz force sits in the antisymmetric off-diagonal components of the diffusion tensor. It

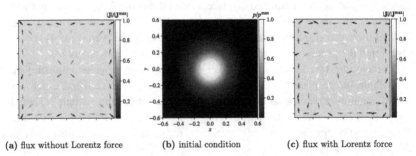

(a) flux without Lorentz force   (b) initial condition   (c) flux with Lorentz force

Figure 1.1 *Probability fluxes with and without Lorentz force.* (b) The scaled probability density p/p^{max} according to a Gaussian initial condition for diffusing particles. (c) When the particles are charged and submitted to a Lorentz force, the probability flux $\mathbf{J}$ (arrows) has the unique characteristic of rotational components. (a) In contrast, in a similar system without the Lorentz force, we observe a pure radial flux. (a) and (c) show the magnitude of the scaled fluxes $|\mathbf{J}|/|\mathbf{J}^{max}|$ after the same reduced time of diffusion

is essentially this property of antisymmetric off-diagonal elements, which together with excluded volume effects gives rise to interesting mathematics and physics.

The idea with which this project was born and which finally resulted in this master thesis dates back to a lunch, taken around March 2020. Because of the Corona pandemic back at that time, cafeterias counted their guests by giving each of them a poker-chip-like disk when entering. Just days ago we became aware of the work by Bruna and Chapman. Discussing our recent work on the effects of the Lorentz force in diffusing systems [15–18] in the group and playing with the poker-chips (i.e. the hard-core interacting particles) on the table, the idea was born, whether we can transfer the unique effect Lorentz force is showing from charged to uncharged particles. The geometric model of Bruna and Chapman seemed to provide a well-suited framework for our ideas.

As one of the applications of the derived theory in this thesis, we will model the motivating example of uncharged Brownian particles diffusing together with charged particles, which therefore feel the effect of the applied magnetic field. But the main results of this work are the predicted collective and self-diffusion coefficients of the Brownian particles under Lorentz force.

Collective diffusion is a measure of the diffusion of all particles simultaneously. It is known to increase with including the collisions between hard-core interacting particles in the description [19]. The self-diffusion, in contrast, measures the diffusive behaviour of a single particle as it undergoes collisions with the others. It is generally believed that collisions of particles reduce the self-diffusion coefficient [20, 21]. For an illustration of collective versus self-diffusion see Figure 1.2.

This work predicts that collisions of charged hard-disks under the effect of Lorentz force result in a crossover from the reduction of self-diffusion to an enhancement. This is an exact analytical prediction, which is caused by the special nature of the antisymmetric off-diagonal elements in the diffusion tensor. We verify this prediction by many-body Brownian-dynamics simulations [22] and present a physical mechanism to explain this effect.

The thesis is organized as follows: In Chapter two we present the theory of coarse-graining the joint probabilistic description by treating the hard-core interactions geometrically. In Chapter three the numerical implementation of the obtained diffusion equations is shown, and the theory is tested on the system of charged and uncharged particles jointly diffusing. Chapter four, as the main result of this work, presents the derivation of the collective and self-diffusion coefficients for charged particles submitted to the Lorentz force. Simulation results are presented and we provide a physical mechanism. Because of the result of an enhancement of the self-diffusion due to collisions being so unexpected, in this project we performed an alternative theoretical work on the phenomenon. Chapter five briefly summarizes

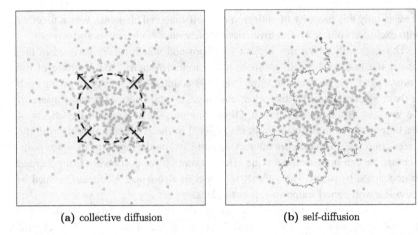

(a) collective diffusion   (b) self-diffusion

Figure 1.2 *Collective and self-diffusion.* The collective diffusion (a) is a phenomenon of many Brownian particles diffusing simultaneously. When they are Gaussian distributed, as in the schematic, collective diffusion can be associated with the growth of the Gaussian variance in time. The self-diffusion (b), in contrast, describes the diffusion of a single Brownian particle interacting with the surrounding particles. Interactions between the particles speed up the collective diffusion of the species, but hinder the self-diffusion of the tagged particle in general

this first-principles approach and shows a remarkable agreement with the afore presented asymptotic theory. We finish Chapter six with a summary and an outlook on future research.

At the date of publishing this thesis, we have already prepared two distinct reports. One mainly covers the asymptotic theory and its results [1], whereas the other publication, which discusses the first-principles approach is being prepared right now.

Theory

2

This chapter[1] presents the theory of including hard-core interactions into the probabilistic description for the time-evolution of one-body distributions of Brownian particles. These particles are subjected to a Lorentz force, which causes an additional rotational effect to the ordinary diffusive flux. We coarse-grain the probabilistic description and arrive at a one-body equation. The contributions of collisions to this equation are built in an integral. Evaluating this with the use of a matched asymptotic expansion, we finish this Chapter by giving a collision-corrected one-body diffusion equation.

2.1 Model Description

2.1.1 Diffusive Motion under Lorentz Force

The presence of an applied magnetic field $\mathbf{B}$ exerts a Lorentz force $q(\mathbf{v}(t) \times \mathbf{B})$ on a particle with charge q and mass m, which moves with velocity $\mathbf{v}(t)$. As apparent, the Lorentz force is a velocity-dependent force, which curves the trajectory of a particle without performing work on it. The stochastic description of this particle, suspended in a solvent reads in presence of the Lorentz force

---

[1]Parts of this Chapter are edited and reprinted with permission from [1]. Copyright (2022) by the American Physical Society.

Supplementary Information The online version contains supplementary material available at https://doi.org/10.1007/978-3-658-39518-6_2.

E. Kalz, *Diffusion under the Effect of Lorentz Force*, BestMasters,
https://doi.org/10.1007/978-3-658-39518-6_2

$$\frac{d\mathbf{x}(t)}{dt} = \mathbf{v}(t), \tag{2.1a}$$

$$m\frac{d\mathbf{v}(t)}{dt} = -m\gamma\ \mathbf{v}(t) + q(\mathbf{v}(t) \times \mathbf{B}) - \mathbf{f}(\mathbf{x}) + \boldsymbol{\xi}(t), \tag{2.1b}$$

where γ is the friction coefficient and $\mathbf{f}(\mathbf{x})$ is an additional external force at the position coordinate $\mathbf{x}(t)$. The effect of the solvent on the particle is modelled by the stochastic vector $\boldsymbol{\xi}(t)$, which is Gaussian distributed with mean $\langle\boldsymbol{\xi}(t)\rangle = \mathbf{0}$ and autocorrelation $\langle\boldsymbol{\xi}(t)\boldsymbol{\xi}^{\mathrm{T}}(t')\rangle = 2k_{\mathrm{B}}T/\gamma\ \delta(t-t')\ 1$. T is the temperature, k_{B} the Boltzmann constant and 1 the identity tensor. $(\cdot)^{\mathrm{T}}$ denotes a matrix transpose. Because of its autocorrelation property, $\boldsymbol{\xi}(t)$ is a white Gaussian noise. Eq.(2.1b) is known as the Langevin equation of motion [23, 24].

One of the properties of the magnetic field is that not all directions in space are equivalent any longer. The magnetic field reduces the diffusion in the plane perpendicular to the field, whereas it leaves the diffusion along the field unaffected. As a result, the diffusion coefficient becomes a tensorial quantity. We are interested in the effect of magnetic field on the diffusive motion, therefore we restrict our analysis for the rest of this thesis to the two-dimensional plane perpendicular to the magnetic field direction $\mathbf{B} = B\hat{\mathbf{e}}_z$, pointing in the auxiliary Cartesian z-direction. For a detailed discussion of the difficulties of applying the theory as presented in this Chapter in three dimensions see the Electronic Supplementary Material, Section B.2.

Obtaining from Eq.(2.1b) the overdamped process, usual methods do fail. They are not able to represent correctly the curving nature of the Lorentz force in the underdamped equation. As Chun et al. [14] have pointed out, the necessary limiting procedures cannot be taken carefully in this methods. As a result, the corresponding probabilistic equations to the (wrong) underdamped equations do not show the characteristic rotational diffusive flux, which is known from simulations to exist [15]. The correct underdamped Langevin equation reads

$$\dot{\mathbf{x}}(t) = -\beta\mathsf{D}\ \mathbf{f}(\mathbf{x}) + \boldsymbol{\eta}(t), \tag{2.2}$$

with $\beta = 1/k_{\mathrm{B}}T$ as the inverse temperature. The procedure to arrive at the underdamped equation is an intricate one. The reason is another property of the magnetic field, the breaking of time-reversal symmetry in the system. The noise $\boldsymbol{\eta}(t)$ is carrying this information in its autocorrelation. It is still a Gaussian $\langle\boldsymbol{\eta}(t)\rangle = \mathbf{0}$, but now a non-white noise $\langle\boldsymbol{\eta}(t)\boldsymbol{\eta}^{\mathrm{T}}(t')\rangle = \mathsf{D}\delta_+(t-t') + \mathsf{D}^{\mathrm{T}}\delta_-(t-t')$. Here Chun et. al introduced variants of the Dirac-delta distribution $\delta_{\pm}(u)$, which are equal to zero for $u \neq 0$, while $\int_0^\infty du\ \delta_+(u) = \int_{-\infty}^0 du\ \delta_-(u) = 1$ and

$\int_0^\infty du\, \delta_-(u) = \int_{-\infty}^0 du\, \delta_+(u) = 0$. These generalized functions reflect the asymmetry in time reversion.

The diffusion tensor D captures the aforementioned isotropy of the two-dimensional space, perpendicular to the direction of the magnetic field, but has antisymmetric off-diagonal elements. They are an artefact of the rotation due to the Lorentz force on a moving charged particle in the underdamped equation (2.1b). The tensor reads

$$D = \frac{D_0}{1+\kappa^2} \begin{pmatrix} 1 & \kappa \\ -\kappa & 1 \end{pmatrix}, \tag{2.3}$$

where $\frac{D_0}{1+\kappa^2}$ is the Lorentz diffusivity, as we tend to call it throughout this thesis to distinguish it from the bare diffusivity D_0. $\kappa = \frac{qB}{\gamma}$ is a measure of the Lorentz force to the drag force, which the Brownian particle experiences.

2.1.2 Probabilistic Description

Because the noise in the overdamped Langevin equation (2.2) for the charged Brownian particle is non-white, usual Itô-calculus cannot be applied to derive a continuity equation for the probability distribution $\mathcal{P}(\mathbf{x}, t)$. The necessary limiting procedures cannot be taken carefully. Iman Abdoli et al. [15] showed that the rotational probability fluxes, which are induced by the Lorentz force, are not reflected in these wrong overdamped equations. But using other techniques, such as the Brinkman method (see for example [25]) or the projection operator method (see for example [26]), the correct equation can be derived to be

$$\frac{\partial}{\partial t}\mathcal{P}(\mathbf{x}, t) = \nabla \cdot D \left[\nabla \mathcal{P}(\mathbf{x}, t) - \beta \mathbf{f}(\mathbf{x})\, \mathcal{P}(\mathbf{x}, t) \right]. \tag{2.4}$$

Note here, that this equation cannot be named as a Fokker-Planck equation, for the Itô-calculus not to apply. We therefore refer to Eq.(2.4) as a continuity equation for the probability density.

One of the goals of this thesis is to go beyond the now understood level of how Lorentz force alters the fundamental stochastic equations. We want to investigate how the probabilistic description is affected, when we allow the particles to interact with each other. As a model, we use the simplest possible interaction between two Brownian particles, the hard-core interaction. Similar to Eq.(2.4), we can also derive the time-evolution equation for the joint marginal probability distribution $P(t) \equiv P(\mathbf{x}_1, \mathbf{x}_2, t)$ of two particles. This reads

$$\frac{\partial}{\partial t} P(t) = \nabla_1 \cdot \mathsf{D}_1 \left[\nabla_1 P(t) - \beta \mathbf{f}_1 \ P(t) \right] + \nabla_2 \cdot \mathsf{D}_2 \left[\nabla_2 P(t) - \beta \mathbf{f}_2 \ P(t) \right], \quad (2.5)$$

where the partial derivatives ∇_i are taken with respect to the ith particles position coordinate $\mathbf{x}_i$. We allow the disk-like particles to differ in the external force $\mathbf{f}_i \equiv \mathbf{f}_i(\mathbf{x}_i)$, their charges q_i, friction coefficients γ_i and hence their bare diffusivities $D_0^{(i)}$, but fix them to be of same diameter σ. This is reflected in the two diffusion tensors $\mathsf{D}_i = \frac{D_0^{(i)}}{1+\kappa^2} \left(\begin{smallmatrix} 1 & \kappa_i \\ -\kappa_i & 1 \end{smallmatrix} \right)$, which both are of the form of Eq.(2.3), but with $\frac{D_0^{(i)}}{1+\kappa_i^2}$ and $\kappa_i = \frac{q_i B}{\gamma_i}$ ($i \in \{1, 2\}$).

Following Bruna and Chapman [10–12], Eq.(2.5) is not defined in whole $\mathbb{R}^2$, but instead in a reduced form $\Omega = \mathbb{R}^2 \times \mathbb{R}^2 \setminus \mathcal{B}$, because of the hard-core interactions between the particles. Here $\mathcal{B} = \{(\mathbf{x}_1, \mathbf{x}_2) \in \mathbb{R}^2 \times \mathbb{R}^2; \ ||\mathbf{x}_1 - \mathbf{x}_2|| \leq \sigma\}$ is the set of all forbidden configurations corresponding to an overlap of the particles. This so-called excluded volume resulting in the reduced configuration space is illustrated in Figure 2.1 (a).

Using this excluded volume, in contrast to common approaches, we treat the hard-core interactions between the Brownian particles in a geometrical sense. The hard-core interactions define a no-flux boundary condition on the so-called collision

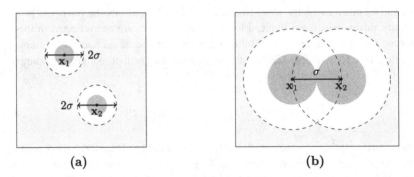

(a)   **(b)**

Figure 2.1 *Excluded volume.* (a) Each Brownian particle (filled disk) carries an excluded volume (white), which is inaccessible to the other particles' centre-of-mass coordinate. Hard-core interactions forbid the centre of the other particle to enter this area. (b) The equation governing the time-evolution of the probability distribution of the two-particle centres is only defined in the light-shaded area, where there is a no-flux boundary condition defined on the collision surface, indicated by the dashed lines. Together the white area forms the forbidden area $\mathcal{B}$. The particles are in contact with each other when the centre-of-mass coordinate of the left particle is on the dashed line of the right particle and vice versa. Figures are reprinted and edited with permission of [1]. Copyright (2022) by the American Physical Society

surface $\partial\Omega = \partial\mathcal{B}$ for the time evolution equation of the joint probability distribution in Eq.(2.5) (see Figure 2.1 (b))

$$D_1 [\nabla_1 P(t) - \beta \mathbf{f}_1 \, P(t)] \cdot \hat{\mathbf{n}}_1 + D_2 [\nabla_2 P(t) - \beta \mathbf{f}_2 \, P(t)] \cdot \hat{\mathbf{n}}_2 = 0, \qquad (2.6)$$

where $\hat{\mathbf{n}}_i = \frac{\mathbf{n}_i}{\|\mathbf{n}_i\|}$ and $\mathbf{n}_i$ is the component of the outward unit normal vector corresponding to the ith particle, i.e. $\hat{\mathbf{n}} = (\mathbf{n}_1, \mathbf{n}_2)$. Note that $\hat{\mathbf{n}}_1 = -\hat{\mathbf{n}}_2$ on $\partial\mathcal{B}$.

We are interested in an effective time-evolution equation on the one-body level for the densities

$$p_1(\mathbf{x}_1, t) = \int_{\Omega(\mathbf{x}_1)} d\mathbf{x}_2 \, P(\mathbf{x}_1, \mathbf{x}_2, t) \qquad (2.7)$$

and

$$p_2(\mathbf{x}_2, t) = \int_{\Omega(\mathbf{x}_2)} d\mathbf{x}_1 \, P(\mathbf{x}_1, \mathbf{x}_2, t) \qquad (2.8)$$

as the one-body densities for particle one and two, respectively. Here $\Omega(\mathbf{x}_i) = \mathbb{R}^2 \backslash B_\sigma(\mathbf{x}_i)$, where $B_\sigma(\mathbf{x}_i)$ is the disk of radius σ around $\mathbf{x}_i$ ($i \in \{1, 2\}$). This reduced configuration space respects that we integrate out one particle, when effectively fixing the other. In this integration we need to take the boundary condition of Eq. (2.6) into account when integrating Eq. (2.5) over the others particles coordinate. For illustration purposes we demonstrate the derivation for the time-evolution equation for the one-body density distribution p_1 of particle one. With an interchange of the particle labels, the time-evolution equation for the other particle can be easily obtained.

The method, which we are adopting here originates from the work of Bruna and Chapman [10–12], where they introduced a way of dealing with excluded volume interactions for reducing the time-evolution equation for a joint probability distribution for ordinary (uncharged) particles through a systematic expansion in the particle diameter σ.

2.2 Diffusion with Finite Size Effects

We want to integrate out the effect of the second particle on the first in the time evolution equation of the probability density

$$\frac{\partial}{\partial t} P(t) = \nabla_1 \cdot D_1 [\nabla_1 P(t) - \beta \mathbf{f}_1 \, P(t)] + \nabla_2 \cdot D_2 [\nabla_2 P(t) - \beta \mathbf{f}_2 \, P(t)]. \qquad (2.9)$$

Therefore we effectively fix particle one and as before denote the reduced configuration space for the coordinate of particle two as $\Omega(\mathbf{x}_1)$. Using the Reynolds transport theorem, we integrate the left-hand side of Eq. (2.5) over $\Omega(\mathbf{x}_1)$. This results in

$$\frac{\partial}{\partial t} \int_{\Omega(\mathbf{x}_1)} d\mathbf{x}_2 \, P(t) = \int_{\Omega(\mathbf{x}_1)} d\mathbf{x}_2 \, \frac{\partial}{\partial t} P(t) + \int_{\partial\Omega(\mathbf{x}_1)} d\mathcal{S}_2 \, (\mathbf{v}^{\Omega(\mathbf{x}_1)} \cdot \hat{\mathbf{n}}_2) P(t), \quad (2.10)$$

where $\mathbf{v}^{\Omega(\mathbf{x}_1)}$ denotes the velocity with which $\Omega(\mathbf{x}_1)$ moves. Note that the projected velocity onto the surface normal is known as surface velocity in the calculus of moving surfaces; the Raynolds transport theorem is a rather common result of in-time moving surfaces (see for example [27]). Since $|\Omega(\mathbf{x}_1)|$ is constant in time, $\mathbf{v}^{\Omega(\mathbf{x}_1)} = \mathbf{0}$. Further we can identify the definition of the one-body probability distribution p_1 from Eq.(2.7) and thus find for the left hand side of integrating Eq.(2.9) over the reduced available volume

$$\int_{\Omega(\mathbf{x}_1)} d\mathbf{x}_2 \, \frac{\partial P(t)}{\partial t} = \frac{\partial p_1(\mathbf{x}_1, t)}{\partial t}. \quad (2.11)$$

The right-hand side of the integrated Eq.(2.9) has two distinct contributions. When integrating over the reduced configuration space $\Omega(\mathbf{x}_1)$, for the second we can apply the divergence theorem, since both, the differentiation and the integration, are with respect to the second particles coordinate

$$\int_{\Omega(\mathbf{x}_1)} d\mathbf{x}_2 \, \nabla_2 \cdot D_2 \left[\nabla_2 P(t) - \beta \mathbf{f}_2 \, P(t)\right] = \int_{\partial B_\sigma(\mathbf{x}_1)} d\mathcal{S}_2 \, \hat{\mathbf{n}}_2 \cdot D_2 \left[\nabla_2 P(t) - \beta \mathbf{f}_2 \, P(t)\right],$$
$$(2.12)$$

where $d\mathcal{S}_2 \, \hat{\mathbf{n}}_2$ is the directed surface element for the excluded volume of particle two and we used that $\partial\Omega(\mathbf{x}_1) = \partial B_\sigma(\mathbf{x}_1)$. Here we can apply the no-flux boundary condition of Eq.(2.6) to obtain

$$\int_{\partial B_\sigma(\mathbf{x}_1)} d\mathcal{S}_2 \, \hat{\mathbf{n}}_2 \cdot D_2 \left[\nabla_2 P(t) - \beta \mathbf{f}_2 \, P(t)\right]$$

$$= -\int_{\partial B_\sigma(\mathbf{x}_1)} d\mathcal{S}_2 \, \hat{\mathbf{n}}_1 \cdot D_1 \left[\nabla_1 P(t) - \beta \mathbf{f}_1 \, P(t)\right]$$
$$(2.13a)$$

$$= \int_{\partial B_\sigma(\mathbf{x}_1)} d\mathcal{S}_2 \, \hat{\mathbf{n}}_2 \cdot D_1 \left[\nabla_1 P(t) - \beta \mathbf{f}_1 \, P(t)\right], \quad$$
$$(2.13b)$$

where we used that $\hat{\mathbf{n}}_2 = -\hat{\mathbf{n}}_1$ on $\partial B_\sigma(\mathbf{x}_1)$ for the last line.

Because for the remaining space-integral the integration variable and the variable of partial differentiation are not of the same particle, the integral cannot be evaluated again with the use of the divergence theorem, but instead is given as

$$\int_{\Omega(\mathbf{x}_1)} d\mathbf{x}_2 \, \nabla_1 \cdot D_1 \left[\nabla_1 P(t) - \beta \mathbf{f}_1 \, P(t) \right] = \nabla_1 \cdot D_1 \left[\nabla_1 p_1(\mathbf{x}_1, t) - \beta \mathbf{f}_1 \, p_1(\mathbf{x}_1, t) \right]$$

$$- \int_{\partial B_\sigma(\mathbf{x}_1)} dS_2 \, \hat{\mathbf{n}}_2 \cdot \left[\left(D_1 + D_1^\mathsf{T} \right) \nabla_1 P(t) + D_1^\mathsf{T} \nabla_2 P(t) - D_1 \beta \mathbf{f}_1 \, P(t) \right]. \quad (2.14)$$

The details of deriving Eq.(2.14) are shown in the Electronic Supplementary Material, Section A.

Collecting results, we obtain an effective equation for the density p_1 of the first particle with the collision effects build in an integral

$$\frac{\partial p_1(\mathbf{x}_1, t)}{\partial t} = \nabla_1 \cdot D_1 \left[\nabla_1 p_1(\mathbf{x}_1, t) - \beta \mathbf{f}_1(\mathbf{x}_1) \, p_1(\mathbf{x}_1, t) \right]$$

$$- \int_{\partial B_\sigma(\mathbf{x}_1)} dS_2 \, \hat{\mathbf{n}}_2 \cdot D_1^\mathsf{T} \left[\nabla_1 P(\mathbf{x}_1, \mathbf{x}_2, t) + \nabla_2 P(\mathbf{x}_1, \mathbf{x}_2, t) \right].$$

$$(2.15)$$

We are left with evaluating the integral living on the collision surface of the two spheres (see again Figure 2.1 (b)), which we refer to as the collision integral. We denote it by

$$I(\mathbf{x}_1, t) = - \int_{\partial B_\sigma(\mathbf{x}_1)} dS_2 \, \hat{\mathbf{n}}_2 \cdot D_1^\mathsf{T} \left[\nabla_1 P(\mathbf{x}_1, \mathbf{x}_2, t) + \nabla_2 P(\mathbf{x}_1, \mathbf{x}_2, t) \right]. \quad (2.16)$$

In the spirit of treating the hard-core collisions between the particles in a geometric sense, Bruna and Chapman [10–12] suggested a method based on matched asymptotic expansion to compute $I(\mathbf{x}_1, t)$ and obtain a closed one-body equation for the time evolution of p_1.

2.3 Matched Asymptotic Expansion

For evaluation of the collision integral in Eq.(2.16), we now change notion to describe the problem. This will respect that the particle interaction is localized near the collision surface $\partial B_\sigma(\mathbf{x}_1)$.

When the two particles are far apart from each other ($\|\mathbf{x}_1 - \mathbf{x}_2\| \gg \sigma$), they can be treated as independent. In this so-called outer region we define $P^{\text{out}}(\mathbf{x}_1, \mathbf{x}_2, t) \equiv$

$P(\mathbf{x}_1, \mathbf{x}_2, t)$ and by the independency argument we have that $P^{\text{out}}(\mathbf{x}_1, \mathbf{x}_2, t) = g_1(\mathbf{x}_1, t)g_2(\mathbf{x}_2, t)$ for some functions g_1 and g_2. Using the normalization condition on P, one can find [10] that indeed $g_1(\mathbf{x}_1, t) = p_1(\mathbf{x}_1, t) + \mathcal{O}(\sigma^2)$ and $g_2(\mathbf{x}_2, t) = p_2(\mathbf{x}_2, t) + \mathcal{O}(\sigma^2)$.

When the two particles are close together ($\|\mathbf{x}_1 - \mathbf{x}_2\| \sim \sigma$), they are correlated and thus, the collision integral will become relevant. In this so-called inner region we assign new coordinates to the problem. We fix the first particle and measure the second with respect to the fixed. Therefore we set $\mathbf{x}_1 \equiv \tilde{\mathbf{x}}_1$ and $\mathbf{x}_2 \equiv \tilde{\mathbf{x}}_1 + \sigma\tilde{\mathbf{x}}$ and define for the inner joint probability $\sigma^2 \tilde{P}(\tilde{\mathbf{x}}_1, \tilde{\mathbf{x}}, t) \equiv P(\mathbf{x}_1, \mathbf{x}_2, t)$, to which we assign also a tilde for convenience. Again we use the short notation $\tilde{P}(t) \equiv \tilde{P}(\tilde{\mathbf{x}}_1, \tilde{\mathbf{x}}, t)$. This coordinate change is illustrated in Figure 2.2.

The reformulated problem of Eq.(2.5) reads in the new coordinates

$$\sigma^2 \frac{\partial \tilde{P}(t)}{\partial t} = \nabla_{\tilde{\mathbf{x}}} \cdot \left[(D_1 + D_2) \, \nabla_{\tilde{\mathbf{x}}} \tilde{P}(t) \right] + \sigma \left[\nabla_{\tilde{\mathbf{x}}} \cdot (D_1 \, \beta \mathbf{f}_1(\tilde{\mathbf{x}}_1) - D_2 \, \beta \mathbf{f}_2(\tilde{\mathbf{x}}_1 + \sigma\tilde{\mathbf{x}})) \, \tilde{P}(t) \right.$$
$$\left. - \nabla_{\tilde{\mathbf{x}}} \cdot \left(D_1 + D_1^{\mathsf{T}} \right) \nabla_{\tilde{\mathbf{x}}_1} \tilde{P}(t) \right] + \sigma^2 \nabla_{\tilde{\mathbf{x}}_1} \cdot D_1 \left[\nabla_{\tilde{\mathbf{x}}_1} \tilde{P}(t) - \beta \mathbf{f}_1(\tilde{\mathbf{x}}_1) \tilde{P}(t) \right], \quad (2.17a)$$

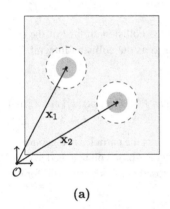   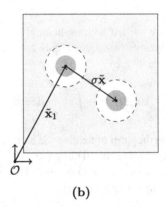

(a)   (b)

Figure 2.2 *Coordinate change.* When the two particles are close in the inner region, i.e. $\|\mathbf{x}_1 - \mathbf{x}_2\| \sim \sigma$ the set of coordinates describing the particles center-coordinate changes from the ordinary (a) to a particle-focused (b). The change of coordinates corresponds to 'fixing' the first particle (previously $\mathbf{x}_1$) and describing the position of the second particle with respect to the position of the first. One among some advantages of this new set of coordinates for the collision problem is that the collision surface now simply is the area of $\|\tilde{\mathbf{x}}\| = 1$. Figures are reprinted and edited with permission of [1]. Copyright (2022) by the American Physical Society

with an obvious choice for $\nabla_{\tilde{x}_1}$ and $\nabla_{\tilde{x}}$ as the partial differential operators with respect to the inner coordinates $\tilde{x}_1$ and $\tilde{x}$, respectively.

In inner coordinates, the collision surface $\partial B_\sigma(x_1)$ is the surface $||\tilde{x}|| = 1$. The boundary condition of Eq.(2.6) in the inner coordinates becomes

$$\tilde{x} \cdot \left[(D_1 + D_2) \, \nabla_{\tilde{x}} \tilde{P}(t) \right] = \sigma \tilde{x} \cdot \left[D_1 \, \nabla_{\tilde{x}_1} \tilde{P}(t) + (D_2 \, \beta f_2(\tilde{x}_1 + \sigma \tilde{x}) - D_1 \, \beta f_1(\tilde{x}_1)) \, \tilde{P}(t) \right],$$
(2.17b)

where we used that $-\hat{n}_2 = \hat{n}_1 = \tilde{x}$ on the collision surface.

In inner variables we need to introduce another boundary condition, since the inner solution $\tilde{P}(t)$ has to match the outer solution $P^{\text{out}}(t)$ as $||\tilde{x}|| \to \infty$. Taylor-expanding the outer solution in powers of the particle diameter σ, one obtains

$$\tilde{P}(\tilde{x}_1, \tilde{x}, t) \sim p_1(\tilde{x}_1, t) \, p_2(\tilde{x}_1, t) + \sigma p_1(\tilde{x}_1, t) \, \tilde{x} \cdot \nabla_{\tilde{x}_1} p_2(\tilde{x}_1, t) + \mathcal{O}(\sigma^2). \quad (2.17c)$$

Eq.(2.17a) together with the boundary condition in Eq.(2.17b) (at $||\tilde{x}|| = 1$) and in Eq.(2.17c) (as $||\tilde{x}|| \to \infty$) defines the two-particle collision problem in inner coordinates. In the spirit of the matched asymptotic expansion we now look for a perturbative solution of this system in powers of the particle diameter σ

$$\tilde{P}(\tilde{x}_1, \tilde{x}, t) = \tilde{P}^{(0)}(\tilde{x}_1, \tilde{x}, t) + \sigma \tilde{P}^{(1)}(\tilde{x}_1, \tilde{x}, t) + \mathcal{O}(\sigma^2). \quad (2.18)$$

Substituting this ansatz into Eq.(2.17a) and arranging in powers of σ yields

$$\sigma^2 \frac{\partial \tilde{P}^{(0)}(t)}{\partial t} + \sigma^3 \frac{\partial \tilde{P}^{(1)}(t)}{\partial t} + \mathcal{O}(\sigma^4) = \nabla_{\tilde{x}} \cdot \left[(D_1 + D_2) \, \nabla_{\tilde{x}} \tilde{P}^{(0)}(t) \right] \quad (2.19a)$$

$$+ \sigma \left[\nabla_{\tilde{x}} \cdot (D_1 + D_2) \, \nabla_{\tilde{x}} \tilde{P}^{(1)}(t) + \nabla_{\tilde{x}} \cdot (D_1 \, \beta f_1(\tilde{x}_1) - D_2 \, \beta f_2(\tilde{x}_1)) \, \tilde{P}^{(0)}(t) \right.$$

$$\left. - \nabla_{\tilde{x}} \cdot \left(D_1 + D_1^T \right) \nabla_{\tilde{x}_1} \tilde{P}^{(0)}(t) \right]$$
(2.19b)

$$+ \sigma^2 \left[\nabla_{\tilde{x}} \cdot (D_1 + D_2) \, \nabla_{\tilde{x}} \tilde{P}^{(2)}(t) + \nabla_{\tilde{x}} \cdot (D_1 \, \beta f_1(\tilde{x}_1) - D_2 \, \beta f_2(\tilde{x}_1)) \, \tilde{P}^{(1)}(t) \right.$$

$$- \nabla_{\tilde{x}} \cdot \left(D_1 + D_1^T \right) \nabla_{\tilde{x}_1} \tilde{P}^{(1)}(t) - \nabla_{\tilde{x}} \cdot D_2 \left(\tilde{x} \cdot \nabla_{\tilde{x}_1} \beta f_2(\tilde{x}_1) \right) \tilde{P}^{(0)}(t)$$

$$\left. + \nabla_{\tilde{x}_1} \cdot D_1 \left(\nabla_{\tilde{x}_1} \tilde{P}^{(0)}(t) - \beta f_1(\tilde{x}_1) \tilde{P}^{(0)}(t) \right) \right] + \mathcal{O}(\sigma^3),$$
(2.19c)

a Laplace-equation for each order, to which we refer as the Laplace-problem of the corresponding order. Note that for collecting terms correctly at first order in

σ, for line (2.19b) we Taylor-expanded the force term, since otherwise this would contribute a hidden order in σ

$$\mathbf{f}_2(\tilde{\mathbf{x}}_1 + \sigma\tilde{\mathbf{x}}) = \mathbf{f}_2(\tilde{\mathbf{x}}_1) + \sigma\tilde{\mathbf{x}} \cdot \nabla_{\tilde{\mathbf{x}}_1}\mathbf{f}_2(\tilde{\mathbf{x}}_1) + \mathcal{O}(\sigma^2). \tag{2.20}$$

The leading order of this Laplace-problem, of which the boundary conditions can be obtained in a similar procedure from Eqs.(2.17b) and (2.17c) is

$$\nabla_{\tilde{\mathbf{x}}} \cdot \nabla_{\tilde{\mathbf{x}}} \tilde{P}^{(0)}(\tilde{\mathbf{x}}_1, \tilde{\mathbf{x}}, t) = 0 \tag{2.21a}$$

$$\tilde{\mathbf{x}} \cdot \left[(\mathsf{D}_1 + \mathsf{D}_2)\, \nabla_{\tilde{\mathbf{x}}} \tilde{P}^{(0)}(\tilde{\mathbf{x}}_1, \tilde{\mathbf{x}}, t) \right] = 0 \qquad \text{on } ||\tilde{\mathbf{x}}|| = 1 \tag{2.21b}$$

$$\tilde{P}^{(0)}(\tilde{\mathbf{x}}_1, \tilde{\mathbf{x}}, t) \sim p_1(\tilde{\mathbf{x}}_1, t)\, p_2(\tilde{\mathbf{x}}_1, t) \qquad \text{as } ||\tilde{\mathbf{x}}|| \to \infty. \tag{2.21c}$$

The solution is straight forward to find and is given by

$$\tilde{P}^{(0)}(\tilde{\mathbf{x}}_1, \tilde{\mathbf{x}}, t) = \tilde{P}^{(0)}(\tilde{\mathbf{x}}_1, t) = p_1(\tilde{\mathbf{x}}_1, t)\, p_2(\tilde{\mathbf{x}}_1, t). \tag{2.22}$$

Including the obtained solution for $\tilde{P}^{(0)}(t)$ into the first order problem in line (2.19b), we find

$$0 = \nabla_{\tilde{\mathbf{x}}} \cdot \left((\mathsf{D}_1 + \mathsf{D}_2)\, \nabla_{\tilde{\mathbf{x}}} \tilde{P}^{(1)}(\tilde{\mathbf{x}}_1, \tilde{\mathbf{x}}, t) \right) + \nabla_{\tilde{\mathbf{x}}} \cdot \left[(\mathsf{D}_1\, \beta\mathbf{f}_1(\tilde{\mathbf{x}}_1) - \mathsf{D}_2\, \beta\mathbf{f}_2(\tilde{\mathbf{x}}_1))\, \tilde{P}^{(0)}(\tilde{\mathbf{x}}_1, t) \right.$$
$$\left. - \nabla_{\tilde{\mathbf{x}}} \cdot \left(\mathsf{D}_1 + \mathsf{D}_1^{\mathsf{T}} \right) \nabla_{\tilde{\mathbf{x}}_1} \tilde{P}^{(0)}(\tilde{\mathbf{x}}_1, t) \right]. \tag{2.23}$$

As a result of Eq.(2.22), $\tilde{P}^{(0)}(\tilde{\mathbf{x}}_1, t)$ is independent of $\tilde{\mathbf{x}}$ and therefore all terms in square brackets are a function of $\tilde{\mathbf{x}}_1$ only. The divergence taken with respect to $\tilde{\mathbf{x}}$ in Eq. (2.23) therefore is zero. Note the necessity of the Taylor-expansion of the external force for this argument to hold.

Finally the problem formulated in Eqs.(2.17a)–(2.17c) at first order in the particle diameter σ reads

$$\nabla_{\tilde{\mathbf{x}}} \cdot \nabla_{\tilde{\mathbf{x}}} \tilde{P}^{(1)}(\tilde{\mathbf{x}}_1, \tilde{\mathbf{x}}, t) = 0 \tag{2.24a}$$

$$\tilde{\mathbf{x}} \cdot \left[(\mathsf{D}_1 + \mathsf{D}_2)\, \nabla_{\tilde{\mathbf{x}}} \tilde{P}^{(1)}(\tilde{\mathbf{x}}_1, \tilde{\mathbf{x}}, t) \right] = \tilde{\mathbf{x}} \cdot \mathbf{A}(\tilde{\mathbf{x}}_1, t) \qquad \text{on } ||\tilde{\mathbf{x}}|| = 1 \tag{2.24b}$$

$$\tilde{P}^{(1)}(\tilde{\mathbf{x}}_1, \tilde{\mathbf{x}}, t) \sim \tilde{\mathbf{x}} \cdot \mathbf{B}(\tilde{\mathbf{x}}_1, t) \qquad \text{as } ||\tilde{\mathbf{x}}|| \to \infty, \tag{2.24c}$$

where

$$A(\tilde{x}_1, t) = D_1 \nabla_{\tilde{x}_1} (p_1(\tilde{x}_1, t) \, p_2(\tilde{x}_1, t)) + (D_2 \, \beta f_2(\tilde{x}_1) - D_1 \, \beta f_1(\tilde{x}_1)) \, p_1(\tilde{x}_1, t) \, p_2(\tilde{x}_1, t),$$
$$\tag{2.25a}$$
$$B(\tilde{x}_1, t) = p_1(\tilde{x}_1, t) \nabla_{\tilde{x}_1} p_2(\tilde{x}_1, t). \tag{2.25b}$$

From a mathematical point of view, solving the given $\tilde{P}^{(1)}$-Laplace-problem, is the most interesting novelty for hard-core interacting charged Brownian particles. But the solution is intricate and therefore its derivation is presented in details in the Electronic Supplementary Material, Section B.1. This solution to the first order problem of Eqs.(2.24a)–(2.24a) reads

$$\tilde{P}^{(1)}(\tilde{x}_1, \tilde{x}, t) = \tilde{x} \cdot B(\tilde{x}_1, t) + \frac{(D_1 + D_2)^T}{\det(D_1 + D_2)} \frac{\tilde{x}}{||\tilde{x}||^2} \cdot \left[(D_1 + D_2) B(\tilde{x}_1, t) - A(\tilde{x}_1, t) \right].$$
$$\tag{2.26}$$

Together with the already derived solution of Eq.(2.22), $\tilde{P}^{(0)}(\tilde{x}_1, t) = p_1(\tilde{x}_1, t)$ $p_2(\tilde{x}_1, t)$, according to the matched asymptotic approximation $\tilde{P} = \tilde{P}^{(0)} + \sigma \tilde{P}^{(1)}$ we find that

$$\tilde{P}(\tilde{x}_1, \tilde{x}, t) = p_1 \, p_2 + \sigma \, p_1 \tilde{x} \cdot \nabla_{\tilde{x}_1} p_2 + \sigma \frac{(D_1 + D_2)^T}{\det(D_1 + D_2)} \frac{\tilde{x}}{||\tilde{x}||^2} \cdot \left[D_2 \, p_1 \nabla_{\tilde{x}_1} p_2 \right.$$
$$\left. - D_1 \, p_2 \nabla_{\tilde{x}_1} p_1 - (D_2 \, \beta f_2 - D_1 \, \beta f_1) \, p_1 \, p_2 \right]. \tag{2.27}$$

This solution is valid up to first order in particle diameter σ. Therefore it captures the ideal particle contribution with the $P^{(0)}$-segment and two particle collisions with the $P^{(1)}$-contribution. Eq.(2.27) is in agreement with the reported result of Bruna and Chapman [10, 12], which considered scalar diffusion. This is a special case of our solution in absence of the magnetic field, since in the case of $\kappa_i = 0$ the diffusion tensors of Eq.(2.3) reduce to $D_i = D_0^{(i)} \mathbb{1}$ ($i \in \{1, 2\}$), the bare diffusion coefficients.

2.4   Collision Integral

We want to use the inner solution, as established in Eq.(2.27), to evaluate the collision integral from Eq.(2.16). In inner coordinates, the integral reads

$$I(\tilde{\mathbf{x}}_1, t) = +\sigma \int_{||\tilde{\mathbf{x}}||=1} d\tilde{\mathcal{S}} \, \tilde{\mathbf{x}} \cdot \left[D_1^T \, \nabla_{\tilde{\mathbf{x}}_1} \, \tilde{P}(t) \right], \tag{2.28}$$

where we used that on the collision surface $||\tilde{\mathbf{x}}|| = 1$ we have that $-\hat{\mathbf{n}}_2 = \hat{\mathbf{n}}_1 = \tilde{\mathbf{x}}$ and hence $d\mathcal{S}_2 \, \hat{\mathbf{n}}_2 = -\sigma \, d\tilde{\mathcal{S}} \, \tilde{\mathbf{x}}$.

Inserting the inner solution, the collision integral reads

$$I_{br}(\tilde{\mathbf{x}}_1, t) = \sigma \, D_1^T \, \nabla_{\tilde{\mathbf{x}}_1} \, (p_1(\tilde{\mathbf{x}}_1, t) \, p_2(\tilde{\mathbf{x}}_1, t)) \cdot \int_{||\tilde{\mathbf{x}}_1||=1} d\tilde{\mathcal{S}} \, \tilde{\mathbf{x}} \tag{2.29a}$$

$$+ \sigma^2 \left[D_1^T \, \nabla_{\tilde{\mathbf{x}}_1} \otimes \left(p_1(\tilde{\mathbf{x}}_1, t) \nabla_{\tilde{\mathbf{x}}_1} p_2(\tilde{\mathbf{x}}_1, t) \right) \right] : \int_{||\tilde{\mathbf{x}}_1||=1} d\tilde{\mathcal{S}} \, \tilde{\mathbf{x}} \otimes \tilde{\mathbf{x}} \tag{2.29b}$$

$$+ \sigma^2 \int_{||\tilde{\mathbf{x}}_1||=1} d\tilde{\mathcal{S}} \, \tilde{\mathbf{x}} \cdot D_1^T \nabla_{\tilde{\mathbf{x}}_1} \left[\frac{(D_1 + D_2)^T}{\det(D_1 + D_2)} \tilde{\mathbf{x}} \cdot D_2 \, p_1(\tilde{\mathbf{x}}_1, t) \nabla_{\tilde{\mathbf{x}}_1} p_2(\tilde{\mathbf{x}}_1, t) \right] \tag{2.29c}$$

$$- \sigma^2 \int_{||\tilde{\mathbf{x}}_1||=1} d\tilde{\mathcal{S}} \, \tilde{\mathbf{x}} \cdot D_1^T \nabla_{\tilde{\mathbf{x}}_1} \left[\frac{(D_1 + D_2)^T}{\det(D_1 + D_2)} \tilde{\mathbf{x}} \cdot D_1 \, p_2(\tilde{\mathbf{x}}_1, t) \nabla_{\tilde{\mathbf{x}}_1} p_1(\tilde{\mathbf{x}}_1, t) \right] \tag{2.29d}$$

$$- \sigma^2 \int_{||\tilde{\mathbf{x}}_1||=1} d\tilde{\mathcal{S}} \, \tilde{\mathbf{x}} \cdot D_1^T \nabla_{\tilde{\mathbf{x}}_1} \left[\frac{(D_1 + D_2)^T}{\det(D_1 + D_2)} \tilde{\mathbf{x}} \cdot \left(D_2 \, \beta \mathbf{f}_2(\tilde{\mathbf{x}}_1) \right. \right.$$
$$\left. \left. - D_1 \, \beta \mathbf{f}_1(\tilde{\mathbf{x}}_1) \right) p_1(\tilde{\mathbf{x}}_1, t) \, p_2(\tilde{\mathbf{x}}_1, t) \right], \tag{2.29e}$$

where $\mathbf{a} \otimes \mathbf{b}$ denotes the outer product between two vectors $\mathbf{a}$, $\mathbf{b}$ and $A : B$ the double contraction between two tensors A, B. Note that $\frac{1}{||\tilde{\mathbf{x}}||^2} = 1$ on the collision surface $||\tilde{\mathbf{x}}|| = 1$.

In the first integral in Eq.(2.29a), corresponding to the $\tilde{P}^{(0)}$-contribution to the collision integral, we are left with performing an integral of the outward normal vector on the whole unit-sphere. With this geometrical insight, the integral is exactly zero. Hence, the zero-order solution does not contribute to the collision integral, which is in agreement with intuition, since it corresponds to ideal point-particles. For the remaining four integrals in lines (2.29b)–(2.29e) the following integral is used, which is valid in two dimensions

$$\int_{||\tilde{\mathbf{x}}||=1} d\tilde{\mathcal{S}} \, \tilde{\mathbf{x}} \otimes \tilde{\mathbf{x}} = \pi \mathbf{1}. \tag{2.30}$$

Defining

$$\beta_i \equiv 1 + \frac{D_1 + D_2}{\det(D_1 + D_2)} D_j, \qquad \gamma_i \equiv \frac{D_1 + D_2}{\det(D_1 + D_2)} D_i, \tag{2.31}$$

for $i = 1, j = 2$ and vice versa, the evaluated collision integral reads

$$I(\tilde{\mathbf{x}}_1, t) = \pi\sigma^2 \, \nabla_{\tilde{\mathbf{x}}_1} \cdot D_1 \big[\boldsymbol{\beta}_1 \, p_1(\tilde{\mathbf{x}}_1, t)\nabla_{\tilde{\mathbf{x}}_1} p_2(\tilde{\mathbf{x}}_1, t) - \boldsymbol{\gamma}_1 \, p_2(\tilde{\mathbf{x}}_1, t)\nabla_{\tilde{\mathbf{x}}_1} p_1(\tilde{\mathbf{x}}_1, t)$$
$$- \big(\boldsymbol{\gamma}_2 \, \beta\mathbf{f}_2(\tilde{\mathbf{x}}_1) - \boldsymbol{\gamma}_1 \, \beta\mathbf{f}_1(\tilde{\mathbf{x}}_1)\big) \, p_1(\tilde{\mathbf{x}}_1, t) \, p_2(\tilde{\mathbf{x}}_1, t)\big]. \quad (2.32)$$

The back-transformation into the original coordinates $\mathbf{x}_1$ and $\mathbf{x}_2$ is straight forward, since the variable $\tilde{\mathbf{x}}$, representing the separation of the two particles, does not appear anymore in Eq.(2.32). This is a self-consistency check since we integrated out the effect of the second particle on the fixed first one. With this, we have expressed the one-body equation for particle one, as desired, solely in terms of one-body distributions.

2.5   Single Species Model

The presented model captures the effect of two Brownian particles of different species diffusing under the effect of Lorentz force and interacting with each other via hard-core interactions. The two particles can differ in the external forces they experience and also in charge, friction and hence their bare diffusivity. But we can also use the presented model to derive the time-evolution equation for the one-body probability density of two particles of the same species, i.e. $p_1 = p_2 \equiv p$ and $\mathbf{f}_1 = \mathbf{f}_2 \equiv \mathbf{f}$. The particles still diffuse under the effect of Lorentz force, but of course have the same diffusion tensor $D_1 = D_2 \equiv D$. The time-evolution equation for p can be written straightforwardly from Eq.(2.15) as

$$\frac{\partial p(\mathbf{x}_1, t)}{\partial t} = \nabla_1 \cdot D \, [\nabla_1 p(\mathbf{x}_1, t) - \beta\mathbf{f}(\mathbf{x}_1) \, p(\mathbf{x}_1, t)]$$
$$- \int_{\partial B_\sigma(\mathbf{x}_1)} dS_2 \, \hat{\mathbf{n}}_2 \cdot D^{\mathsf{T}} \, [\nabla_1 P(\mathbf{x}_1, \mathbf{x}_2, t) + \nabla_2 P(\mathbf{x}_1, \mathbf{x}_2, t)]. \quad (2.33)$$

The collision integral can be rewritten into inner coordinates in a similar way as in the two-species model, but reduces tremendously since all inter-species effects vanish. Eq.(2.32) in this single species case reads

$$I(\tilde{\mathbf{x}}_1, t) = \pi\sigma^2 \, \nabla_{\tilde{\mathbf{x}}_1} \cdot D \, \big[p(\tilde{\mathbf{x}}_1, t)\nabla_{\tilde{\mathbf{x}}_1} p(\tilde{\mathbf{x}}_1, t) \big]. \quad (2.34)$$

As one would expect, the variable $\tilde{\mathbf{x}}$ as representing the second particle vanishes at this order and we simply can back-transform into the original variables $\mathbf{x}_1$ and $\mathbf{x}_2$.

The closed equation on the one-body density level reads

$$\frac{\partial p(\mathbf{x}_1, t)}{\partial t} = \nabla_1 \cdot \mathsf{D} \left[\nabla_1 p(\mathbf{x}_1, t) - \beta \mathbf{f}(\mathbf{x}_1) \, p(\mathbf{x}_1, t) + \pi \sigma^2 \, p(\mathbf{x}_1, t) \nabla_1 p(\mathbf{x}_1, t) \right].$$
(2.35)

Considering the special case of absent external force in the upper diffusion equation, i.e. $\mathbf{f}(\mathbf{x}_1) = \mathbf{0}$, we can highlight one mathematical subtlety of this diffusion equation. Eq.(2.35) now is invariant under replacing the diffusion tensor D by its transposed D^T. Due to the conservation of probability, the time evolution equation for the density $\partial_t p(\mathbf{x}_1, t) = -\nabla \cdot \mathbf{J}(\mathbf{x}_1, t)$ has the form of a continuity equation. And thus, the probability flux $\mathbf{J}(\mathbf{x}_1, t)$ is indeterminate on a curl, which can be added. The antisymmetric elements of the diffusion tensor D exactly represent such a curl

$$\mathbf{J}(\mathbf{x}_1, t) = -\mathsf{D} \nabla p(\mathbf{x}_1, t) = -\frac{D_0}{1 + \kappa^2} \left[\nabla p(\mathbf{x}_1, t) + \kappa \hat{\mathbf{n}}_B \times \nabla p(\mathbf{x}_1, t) \right], \quad (2.36)$$

where $\hat{\mathbf{n}}_B$ is the normal vector of the magnetic field $\mathbf{B}$, which points along the auxiliary z-direction. Now having D^T instead of D changes the sign of the antisymmetric off-diagonal elements, which then effectively changes the sign of the curl, on which the probability flux is indeterminate.

We can resolve this problem by recalling the point-particle limit ($\sigma = 0$) of Eq.(2.35), i.e.

$$\frac{\partial p(\mathbf{x}_1, t)}{\partial t} = \nabla_{\mathbf{x}_1} \cdot \left[\mathsf{D} \nabla_{\mathbf{x}_1} \, p(\mathbf{x}_1, t) - \beta \mathsf{D} \, \mathbf{f}(\mathbf{x}_1) p(\mathbf{x}_1, t) \right]. \tag{2.37}$$

A magnetic field performs no work on a system in equilibrium, hence the steady state ($\partial_t p = 0$) cannot show any fingerprints of the magnetic field. We should be able to factor out the magnetic field contribution. This is only possible, when the two diffusion matrices multiplying the diffusion gradient and the drift term are equal

$$0 = \nabla \cdot \mathsf{D} \underbrace{\left[\nabla p(\mathbf{x}_1, t) - \beta \mathbf{f}(\mathbf{x}_1) \, p(\mathbf{x}_1, t) \right]}_{\text{independent of } \mathbf{B}}, \tag{2.38}$$

and hence, the diffusion gradient has to have the same multiplying diffusion tensor as the drift term has, D. Exactly the same argument holds with the $p\nabla p$-term in Eq.(2.35), since it can be rewritten as $\nabla p^2/2$. The steady-state solution has to be independent of the magnetic field, therefore only D is possible as a common diffusion tensor. The far-reaching consequence of this mathematical subtlety will become evident when we derive the self-diffusion coefficient for the charged system in Section 4.1.

The extension from two particles in the system towards N can be implemented straight forward at the order of $\mathcal{O}(\sigma^2)$. Since at this order only pairwise particle interactions are relevant, we know that one particle has $(N - 1)$ inner regions, one with each of the remaining particles. Thus we can write

$$\frac{\partial p(\mathbf{x}_1, t)}{\partial t} = \nabla_1 \cdot \mathsf{D} \left[\nabla_1 p(\mathbf{x}_1, t) - \beta \mathbf{f}(\mathbf{x}_1) \, p(\mathbf{x}_1, t) + (N - 1)\pi\sigma^2 \, p(\mathbf{x}_1, t)\nabla p(\mathbf{x}_1, t) \right].$$
(2.39)

We can see in this equation, that the hard-core interactions lead to a modified diffusion coefficient. The additional term is proportional to the excluded volume $\frac{\pi\sigma^2}{4}$. The equation is consistent with the work by Bruna and Chapman [12] on scalar diffusion, i.e. $\mathsf{D} = D_0 \mathbf{1}$ and also earlier work on collective diffusion by Felderhof [13]. Whereas Bruna and Chapman generalized the work of Felderhof to situations, where p is not necessarily close to uniform, we generalized it to non-scalar diffusion tensors. As it will become apparent later, this leads to non-trivial predictions for the diffusion coefficients of such a species.

2.6   Final Equation

By considering the single-species version of our theory in the preceding Section, we were drawn aside from the main route for one moment. It will shortly become evident, why this detour was necessary at this point. Now we are concerned with including the result of the evaluated collision integral of Eq.(2.32) into the defining one-body equation for the density of particle one. Since we arrived at an equation on the one-body level, we only need one variable for description. Therefore we abbreviate $p_1 \equiv p_1(\mathbf{x}, t)$ and $p_2 \equiv p_2(\mathbf{x}, t)$. Furthermore we denote partial derivatives with respect to this coordinate $\mathbf{x}$ simply as ∇. The time-evolution equation for p_1 thus reads

$$\frac{\partial p_1(\mathbf{x}, t)}{\partial t} = \nabla \cdot \mathsf{D}_1 \left[\nabla p_1 - \beta \mathbf{f}_1(\mathbf{x}) \, p_1 + \pi\sigma^2 \left(\boldsymbol{\beta}_1 \, p_1 \nabla p_2 - \boldsymbol{\gamma}_1 \, p_2 \nabla p_1 \right) \right.$$
$$\left. - \pi\sigma^2 \left(\boldsymbol{\gamma}_2 \, \beta \mathbf{f}_2(\mathbf{x}) - \boldsymbol{\gamma}_1 \, \beta \mathbf{f}_1(\mathbf{x}) \right) p_1 \, p_2 \right]. \quad (2.40)$$

The extension from one to N_2 particles of species two is straight forward up to order $\mathcal{O}(\sigma^2)$, since at this order only pairwise particle interactions need to be considered. For N_2 arbitrary, the particle of species one has N_2 inner regions with particles of species two, one with each of the N_2 particles. Hence there are N_2 copies of the corresponding non-linear term in Eq.(2.40). Similarly for N_1 arbitrary. The particle in focus can have $N_1 - 1$ pairwise interactions with the remaining particles of its own species. As yet the model does not capture these intra-species contributions, we have to recall the non-linear term from the single-species model in Eq.(2.39) to which derivation we devoted the last Section. The time-evolution equation for the one-body distribution of a particle of species one on a level of pairwise particle interactions thus reads

$$\frac{\partial p_1(\mathbf{x}, t)}{\partial t} = \nabla \cdot \mathsf{D}_1 \left[\nabla p_1 - \beta \mathbf{f}_1(\mathbf{x})\, p_1 \right.$$

$$+ (N_1 - 1)\sigma^2 \pi \; p_1 \nabla p_1 + N_2 \pi \sigma^2 \left(\boldsymbol{\beta}_1 \; p_1 \nabla p_2 - \boldsymbol{\gamma}_1 \; p_2 \nabla p_1 \right)$$

$$\left. - N_2 \pi \sigma^2 \left(\boldsymbol{\gamma}_2 \; \beta \mathbf{f}_2(\mathbf{x}) - \boldsymbol{\gamma}_1 \; \beta \mathbf{f}_1(\mathbf{x}) \right) p_1 \; p_2 \right]. \qquad (2.41)$$

In the case of scalar diffusion, i.e. $\mathsf{D}_i = D_0^{(i)}$ and hence $\boldsymbol{\beta}_i = \beta_i \mathbf{1}$ and $\boldsymbol{\gamma}_i = \gamma_i \mathbf{1}$ for $i \in \{1, 2\}$, Bruna and Chapman could give a clear physical interpretation for the effect of collisions of particles of species one with species two in the terms proportional to $+\beta_1 \; p_1 \nabla p_2$ and $-\gamma_1 \; p_2 \nabla p_1$ [12]. The first term is a drift due to the biasing of the random walk of species one in the presence of a gradient of species two and the second term represents the reduction of the diffusion due to collisions of particles with the other species. Incorporating collisions with species two into the time-evolution equation for species one results in an interplay of these different effects. Moreover, the diffusion of species one is enhanced by collisions with other particles of the same species (the term proportional to $+p_1 \nabla p_1$). Such an interpretation in the case of tensorial diffusion is unfortunately hard to find. The coefficients $\boldsymbol{\beta}_1$ and $\boldsymbol{\gamma}_1$ in the equation for species one have the characteristic anti-symmetric off-diagonal form of Eq.(2.3). Therefore it is not clear if enhancement or reduction of diffusion can be unambiguously attributed to these terms since the components are mixed up in both of the terms $+\boldsymbol{\beta}_1 \; p_1 \nabla p_2$ and $-\boldsymbol{\gamma}_1 \; p_2 \nabla p_1$.

The time-evolution equation for p_2 can be read-off from Eq.(2.41) by an interchange of particle-labels to be

$$\frac{\partial p_2(\mathbf{x}, t)}{\partial t} = \nabla \cdot \mathsf{D}_2 \left[\nabla p_2 - \beta \mathbf{f}_2(\mathbf{x})\, p_2 \right.$$

$$+ (N_2 - 1)\sigma^2 \pi \; p_2 \nabla p_2 + N_1 \pi \sigma^2 \left(\boldsymbol{\beta}_2 \; p_2 \nabla p_1 - \boldsymbol{\gamma}_2 \; p_1 \nabla p_2 \right)$$

$$\left. + N_1 \pi \sigma^2 \left(\boldsymbol{\gamma}_1 \; \beta \mathbf{f}_1(\mathbf{x}) - \boldsymbol{\gamma}_2 \; \beta \mathbf{f}_2(\mathbf{x}) \right) p_2 \; p_1 \right]. \qquad (2.42)$$

The one-body time evolution equation for species one and two of Eqs.(2.41) and (2.42) together may be written in a joint matrix-form

$$\frac{\partial}{\partial t}\begin{pmatrix} p_1(\mathbf{x},t) \\ p_2(\mathbf{x},t) \end{pmatrix} = \begin{pmatrix} \nabla \\ \nabla \end{pmatrix} \cdot \left[\mathbb{D}(p_1,p_2)\begin{pmatrix} \nabla p_1(\mathbf{x},t) \\ \nabla p_2(\mathbf{x},t) \end{pmatrix} - \mathbb{F}(p_1,p_2)\begin{pmatrix} p_1(\mathbf{x},t) \\ p_2(\mathbf{x},t) \end{pmatrix} \right],$$
(2.43)

where

$$\mathbb{D} = \begin{bmatrix} D_1\left(1+(N_1-1)\sigma^2\pi\, p_1 - N_2\sigma^2\pi\, \gamma_1\, p_2\right) & N_2 D_1\sigma^2\pi\, \beta_1\, p_1 \\ N_1 D_2\sigma^2\pi\, \beta_2\, p_2 & D_2\left(1+(N_2-1)\sigma^2\pi\, p_2 - N_1\sigma^2\pi\gamma_2\, p_1\right) \end{bmatrix}$$
(2.44)

is the diffusion matrix and

$$\mathbb{F} = \beta\begin{bmatrix} D_1\, \mathbf{f}_1(\mathbf{x}) & N_2\pi\sigma^2\left(\gamma_2\, \mathbf{f}_2(\mathbf{x}) - \gamma_1\, \mathbf{f}_1(\mathbf{x})\right)p_1 \\ N_1\pi\sigma^2\left(\gamma_1\, \mathbf{f}_1(\mathbf{x}) - \gamma_2\, \mathbf{f}_2(\mathbf{x})\right)p_2 & D_2\, \mathbf{f}_2(\mathbf{x}) \end{bmatrix}$$
(2.45)

is the drift matrix. Both are local in time and space, due to their dependence on $p_1(\mathbf{x},t)$ and $p_2(\mathbf{x},t)$. Note the double use of the symbol β, once as the inverse temperature, mutliplying the drift matrix and once as an inter-species coefficient in the diffusion matrix. These coefficients β_i and γ_i measuring the inter-species effects are given in Eq.(2.31). They read

$$\beta_i = 1 + \frac{D_1 + D_2}{\det(D_1 + D_2)}D_j, \qquad \gamma_i = \frac{D_1 + D_2}{\det(D_1 + D_2)}D_i$$

for $i = 1, j = 2$ and vice versa.

2.7 Summary

In this Chapter, we presented the theory of including hard-core interactions between charged particles under the effect of Lorentz force into the probabilistic description in terms of one-body densities. Therefore we followed a geometrical approach, first presented by Bruna and Chapman [10–12], who considered the interactions as an excluded volume each particle is carrying concerning the others. By using this method, when coarse-graining the equation for the joint probability distribution, the configuration space to be integrated over is reduced. This resulted in an integral being defined wherever the singular interaction is present, at the surface of the particles. This integral captured the two-body collision effects and could be evaluated when changing to a particle-centred frame of reference. The perturbative ansatz, which we made to solve the integral was truncated after two-body collision contributions, whereas solving the two-body problem was the true mathematical novelty of this work. When generalizing two particles in the system to an arbitrary number, from the present theory we obtained the inter-species correction terms due to collisions. But for intra-species corrections, we reduced the model to a single-species theory to include the corresponding terms. Finally, we arrived at two diffusion equations on the one-body level for particles of the different species, with included intra- and inter-species collision corrections due to hard-core interactions.

Numerical Results

3

In this Chapter, we present the numerical implementation of the diffusion equation for the two species. Numerical subtleties including the special nature of no-flux boundaries under the effect of Lorentz force are presented. Derivatives in x- and y-direction are coupled, which leads to a new type of boundary condition. We apply the theory to one of the motivating questions of this work; charged and uncharged particles diffusing together under the effect of Lorentz force. As one might expect, the uncharged particles pick up the unique rotational effect of the charged particles due to hard-core interactions between the particles.

3.1 Numerical Method

Charged particles under the effect of Lorentz force sense a boundary differently than uncharged particles do. This difference is depicted in Figure 3.1. There we show the probability distribution and flux for charged particles subjected to Lorentz force in (a) and uncharged particles in (b). They are obtained from numerically solving the time evolution equations, which we derived in the preceding Chapter. The reason for the apparent different behaviour is that a no-flux boundary condition, originating from particle-conservation, in terms of a magnetic field is different from an ordinary von-Neumann condition. Given the diffusion equation $\frac{\partial}{\partial t} p(\mathbf{x}, t) = \nabla \cdot \mathbf{D} \nabla p(\mathbf{x}, t)$ for the one-body density distribution $p(\mathbf{x}, t)$ of ideal charged particles under the effect of Lorentz force, the boundary condition reads

$$\mathbf{b} \cdot \nabla p(\mathbf{x}, t) = 0, \tag{3.1}$$

in the literature also known as an oblique boundary condition [28]. In case of an impenetrable wall in x-direction, the oblique vector reads $\mathbf{b} = \left(\begin{smallmatrix} 1 \\ \kappa \end{smallmatrix} \right)$. Throughout

E. Kalz, *Diffusion under the Effect of Lorentz Force*, BestMasters, https://doi.org/10.1007/978-3-658-39518-6_3

this Chapter we restrict the analysis to this impenetrable wall in x-direction as an example for the effect of boundaries. The origin of this condition, which is different from an ordinary von-Neumann condition $\mathbf{b} \cdot \nabla p(\mathbf{x}, t) = 0$, where $\mathbf{b} = \hat{\mathbf{n}}_x = \left(\begin{smallmatrix} 1 \\ 0 \end{smallmatrix}\right)$, is the presence of the antisymmetric off-diagonal elements in the diffusion tensor.

The diffusion equation as a continuity equation suggests using a forward Euler implementation for the time iteration. In the following equations $_{(i,j)}p^n$ is the probability density at time-step n and grid-point (i, j) on the two-dimensional Cartesian grid. $_{(i,j)}p^{n+1}$ in the forward Euler scheme then can be calculated as

$$_{(i,j)}p^{n+1} = {}_{(i,j)}p^n + \Delta t \left(\partial_x \, _{(i,j)}J_x^n + \partial_y \, _{(i,j)}J_y^n \right), \tag{3.2}$$

where $_{(i,j)}J_x^n$ is the probability flux in x-direction at time-step n and grid-point (i, j), similarly $_{(i,j)}J_y^n$ the y-flux. ∂_x and ∂_y are discrete spatial derivatives in the x- and y-directions, respectively. Using the forward Euler scheme restricts the time-step to $\Delta t \leq \frac{\Delta x^2}{2D_0}(1 + \kappa^2)$, the so called von-Neumann stability criterion [29]. Here $\Delta x = \Delta y$ is the discretization of the grid and $\frac{D_0}{1+\kappa^2}$ the Lorentz diffusivity of the particles.

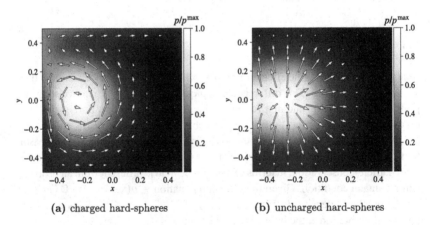

(a) charged hard-spheres   (b) uncharged hard-spheres

Figure 3.1 *Charged and uncharged particles sensing a boundary.* The scaled one-body probability distribution p/p^{max} corresponding to $N = 20$ hard-spheres of diameter $\sigma/L = 0.01$, where L is the system size, are initially Gaussian distributed (center: $(-0.2, 0)$, mean: 0, variance: 0.1 in rescaled units). Shown are two snapshots at the same reduced time of diffusion $\tilde{t} = \frac{D_0}{(1+\kappa^2)L^2}t = 0.01$ for two species of same (Lorentz) diffusivity $\frac{D_0}{1+\kappa^2}$. (a) shows charged hard-disks ($\kappa = 5$) and (b) shows uncharged hard-disks ($\kappa = 0$). The magnetic field adds on top of the radially outward diffusive flux a rotational flux. The fluxes are shown as white arrows with their length indicating the strength

3.2 Central Derivative

Using a central spatial scheme, instead of a forward or backward spatial scheme to compute the derivative on the grid, allows implementation of the oblique no-flux boundary condition of Eq.(3.1) at the impenetrable wall in x-direction in a straightforward way. The central derivative is a scheme with a second-order truncation error, which can also be viewed as an average of two forward methods

$$\partial_x^c{}_{(i,j)}p^n = \frac{(i+1,j)p^n - (i-1,j)p^n}{2\Delta x} \tag{3.3a}$$

$$= \frac{1}{2}\left(\frac{(i+1,j)p^n - (i,j)p^n}{\Delta x} + \frac{(i,j)p^n - (i-1,j)p^n}{\Delta x}\right) \tag{3.3b}$$

$$= \frac{1}{2}\left(\partial_x^{f,out}{}_{(i,j)}p^n + \partial_x^{f,in}{}_{(i,j)}p^n\right), \tag{3.3c}$$

where ∂_x^c is the central derivative operator in x-direction and ∂_x^f the forward operator.

As Eq.(3.3c) suggests, the central derivative averages over the incoming and outgoing derivatives in x-direction, which are given by the incoming forward derivative $\partial_x^{f,in}$ and the outgoing forward derivative $\partial_x^{f,out}$. For an illustration of a central versus a forward derivative see Figure 3.2. The major difference, which is emphasized in this visualization is the position inside the cell, to which each derivative scheme is assigned to. Whereas a central derivative lives at the centre of mass of a cell, a forward derivative is assigned to the edge between two neighbouring cells. A boundary condition, such as an impenetrable wall, is also assigned to the edge between two neighbouring cells. Hence the forward derivative here has to be zero. This can be implemented by using a ghost cell condition $(i_{max}+1,j)p^n = (i_{max},j)p^n$ and is done in the same way in the central scheme. Applying the condition, the forward outward derivative in Eq.(3.3c) is set to zero. Note that the factor $1/2$ accounts for the linear interpolation between the edge points.

But despite the better truncation error of central derivatives, there is a more fundamental aspect of using the central derivative scheme over the forward scheme. To form new quantities like the x-probability-flux, the diffusion tensor couples the derivatives in x and in y-direction due to the magnetic field. This probability flux in x-direction $(i,j)J_x^n$ at any grid point (i,j) in a system under Lorentz force is given as $(i,j)J_x^n = \partial_x{}_{(i,j)}p^n + \kappa\partial_y{}_{(i,j)}p^n$ and reads with the use of central derivatives

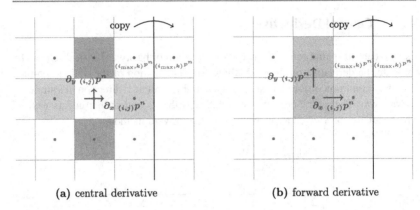

(a) central derivative   **(b)** forward derivative

Figure 3.2 *Central versus forward derivative.* The difference between a central derivative (a) and a forward derivative (b) for a numerical implementation of the gradient operator on a space grid is illustrated schematically. Calculating the spatial derivative using the central scheme, the two next-neighbouring old densities are used, and the derivative lives in the centre of the cell. In implementing the forward scheme, one uses the old densities from the same cell and its forward neighbour in direction of the derivative, which now lives on the boundary between the two cells. The ghost-cell method is applied to represent the no-flux boundary at an impenetrable wall (thick line)

$$(i,j) J_x^n = \quad \partial_x^c \, (i,j) P^n \quad + \quad \kappa \, \partial_y^c \, (i,j) P^n \tag{3.4a}$$

$$= \frac{1}{2} \left(\partial_x^{f,out} \, (i,j) P^n + \partial_x^{f,in} \, (i,j) P^n \right) + \frac{\kappa}{2} \left(\partial_y^{f,out} \, (i,j) P^n + \partial_y^{f,in} \, (i,j) P^n \right), \tag{3.4b}$$

as it also can be expressed with the use of forward derivatives. The coupling of x- and y-derivatives now requires a discrete scheme, which treats both derivatives equal concerning the position in the cell, to which they are assigned to. As already mentioned, the central derivative operator lives in the middle of the cell, whereas the forward operator lives on the boundary between two neighbouring cells. Updating each density cell without the necessity of assigning a new scheme for all boundary-layered cells, the central scheme is the appropriate scheme for the apparent symmetry.

But using the central derivative for the whole grid also intricates numerical procedures in one aspect. As we already argued, the impenetrability condition of boundary requires the outgoing forward derivative to be zero (i.e. no leaking particles) and to ensure this, we apply at the last step at each time-iteration the ghost-cell condition for the density cells. But since the diffusion equation is a second-order differential equation in space, the central derivative scheme has to be applied twice. Hence, when differentiating the flux, we need to provide another ghost-cell condition for the flux-derivatives, a zero-flux ghost-cell. This means for example in x-direction: $_{(i_{max}+1,j)}J_x^n = 0$.

Now that we have described the numerical implementation in sufficient detail, we can use the presented formalism to comment on one interesting subtlety of the magnetic field. Updating the density cells according to the forward Euler scheme of Eq.(3.2) in the bulk we find

$$_{(i,j)}p^{n+1} = \,_{(i,j)}p^n + \frac{\Delta t}{2\Delta x}\left(\partial_x^c \,_{(i+1,j)}p^n - \partial_x^c \,_{(i-1,j)}p^n + \partial_y^c \,_{(i,j+1)}p^n - \partial_y^c \,_{(i,j-1)}p^n\right).$$
(3.5)

There is no dependence on κ further than in the time-step, which has to be chosen according to the already mentioned von-Neumann stability criterion $\Delta t \leq \frac{\Delta x^2}{2D_0}(1 + \kappa^2)$ and $\frac{D_0}{1+\kappa^2}$ as the Lorentz diffusivity. The same numerical behaviour as in Eq.(3.5) would be achieved for the density of an uncharged species with a rescaled bare diffusivity D_0. Note that for simplicity of the argument we took an equal discretization in x- and y-direction $\Delta x = \Delta y$.

The situation changes when the density is updated directly at the impenetrable wall. Here the forward Euler scheme of Eq.(3.2) reads

$$_{(i,j)}p^{n+1} = \,_{(i,j)}p^n + \frac{\Delta t}{2\Delta x}\left(-\partial_x^c \,_{(i-1,j)}p^n + \partial_y^c \,_{(i,j+1)}p^n - \partial_y^c \,_{(i,j-1)}p^n - \kappa\partial_y^c \,_{(i,j)}p^n\right).$$
(3.6)

Apparently, the numerical procedure has a direct fingerprint of the magnetic field parameter κ, induced by the presence of a boundary and with no counterpart in a system without a Lorentz force. The difference of updating the density in the bulk as in Eq.(3.5) and at the boundary as in Eq.(3.6) is schematically illustrated in Figure 3.3. For a visual impression of the result of this magnetic-induced different sensing of a boundary in comparison to non-charged particles, see again Figure 3.1 in the preceding Section.

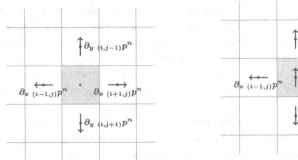

(a) contribution to density-update in bulk

(b) contribution to density-update at a boundary

Figure 3.3 *Numerically updating in the bulk and at a boundary.* The density $(i,j)P^n$ in the shaded cell at grid-point (i, j) is updated according to the forward Euler scheme. As intuition suggests, the probability flux driven density change in the cell can be decomposed into contributions from the densities of neighbouring cells, using the central derivative scheme. (a) shows that updating a cell in the bulk requires symmetric contributions from all next-neighbouring cells, whereas the magnetic field enters only implicitly via the time-step Δt. The numerical solution is indistinguishable from an uncharged species with modified bare diffusivity. The situation changes when the cell is updated at a reflecting boundary, as depicted in (b). The magnetic field enters explicitly into the numerical scheme via κ, a contribution from the cell itself

3.3   Charged versus Uncharged Particles in 2 Dimensions

In this section, we present the first application of the theory. The project, which finally resulted in this master thesis started with the question, of whether it is possible to transfer the unique rotational effect, that charged particles subjected to a Lorentz force are showing, to uncharged particles. We wondered whether this effect can be transmitted via hard-core interactions between the species.

We answer this question by numerically solving Eq.(2.43) with the method presented in this Chapter. We study the diffusion of $N_1 = N_2 = 20$ hard-spheres of diameter $\sigma/L = 0.01$ with L as the system size. Species one is charged ($\kappa_1 = \frac{qB}{\gamma} = 5$), whereas species two is uncharged ($\kappa_2 = 0$). Therefore species two does not experience the effect of the Lorentz force, which an applied magnetic field is inducing. Whereas the charged species has opposite off-diagonal elements in the tensor, i.e. $D_1 = \frac{D_0}{1+\kappa_1^2} \begin{pmatrix} 1 & \kappa_1 \\ -\kappa_1 & 1 \end{pmatrix}$, the uncharged species has a scalar diffusion coefficient, i.e. $D_2 = D_0 1$. Both species are modelled to have the same bare diffusion coefficient D_0. Note that $\phi = 1.3\%$.

To study the effect, which collisions between the charged and uncharged particles are producing, we set up the uncharged species (p_2) initially in equilibrium and the charged species (p_1) to be Gaussian distributed, as it can be seen in the first row of Figure 3.4.

There are now two distinct effects visible in the numerical solution of Eq.(2.43) starting with the initial situation as described. The two species are time-evolving together in a two-dimensional box with reflecting boundaries. Figure 3.4 shows in row two and three two snapshots in time for the two species, taken at reduced time of diffusion $\tilde{t} = \frac{D_0}{(1+\kappa^2)L^2}t$ of $\tilde{t}_1 = 2 \cdot 10^{-2}$ and $\tilde{t}_2 = 2 \cdot 10^{-1}$ for row two and three, respectively. Note, that within one species (one column) the colour bar, as indicating deviations in the density, is taken to be the same, whereas not among the species. The reason is, that the effects are of different magnitude for the charged and uncharged species.

As it is apparent, the uncharged species, initially in equilibrium, is disturbed from its uniform distribution by collisions with the initially Gaussian distributed charged species. The uncharged species is drawing aside from high concentration regions of the charged species. Comparing rows two and three, the depletion of the charged species is correlated to the depletion of the uncharged species. This is intuitively clear as an effect of excluded volume interactions.

The true novelty of charged particles under the effect of Lorentz force is the second visible effect. As it was already shown in Figure 3.1 (a), charged particles under Lorentz force sense the boundary differently, than uncharged particles do. They seem to slip along the boundary, where the handiness of the rotation is determined by the direction of the magnetic field. This different way of sensing the reflecting boundary is also visible in the time-evolution plot of the charged density in row three. Now it is surprising, that the uncharged species is mimicking this property, even though the particles are reflected in a von-Neumann way by a hard wall (see again Figure 3.1 (b)). The inter-species effects of collisions transfer this otherwise unique effect of boundary sensing from the charged to the uncharged species. The reason for this becomes apparent when looking at the probability fluxes. From Figure 3.1 we know that uncharged particles have the typical radial diffusive flux outward from the peak of the distribution, whereas charged particles under Lorentz-force additionally have a rotational flux. The latter also can be seen in the time evolution of the charged species, but in proceeding time, the uncharged species adopts this unique flux property (row three).

To summarize, we were able to transfer the unique property of the Lorentz-flux from a charged species to an uncharged species only by hard-core collisions. By this,

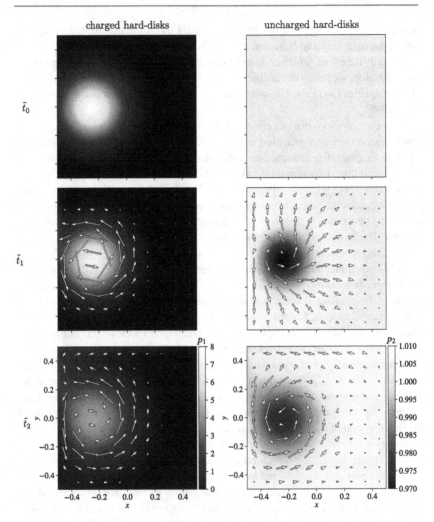

Figure 3.4 *Charged and uncharged particles diffusing with hard-core interactions.* The time evolution of the probability density p of an initially Gaussian distributed (center: $(-0.25, 0)$, mean: 0, variance: 0.1 in rescaled units) charged species of $N_1 = 20$ hard-disks (p_1, left column). They are diffusing in the same system as $N_2 = 20$ uncharged hard-disks (p_2, right column), which are initially in equilibrium. Shown are the three snapshots in reduced diffusion time $\tilde{t} = \frac{D_0}{(1+\kappa^2)L^2}t$, where L is the system size and $\frac{D_0}{1+\kappa^2}$ is the Lorentz diffusivity of the two species, both with identical bare diffusivity D_0. The three snapshots are taken at $\tilde{t}_0 = 0$, $\tilde{t}_1 = 2 \cdot 10^{-2}$ and $\tilde{t}_2 = 2 \cdot 10^{-1}$. The probability fluxes are shown as white arrows, with their length indicating the strength. Note the difference in order of magnitude of the color-coding of charged versus uncharged particles (left versus right column)

we could make the uncharged species sense the reflecting boundary in an oblique way, which usually is unique to Lorentz force, instead of an ordinary von-Neumann way. The far-reaching consequences of this effect will become evident in the next Chapter.

3.4   Summary

In this Chapter, we presented a numerical method to solve the differential equations from Chapter 2. Diffusion under the effect of Lorentz force constitutes a special system to deal with numerically. Specifically, updating a space-discretized system at reflecting boundaries is different from systems without the effect of Lorentz force. The diffusion tensor, in contrast to the ordinary diffusion scalar, couples spatial derivatives of different directions, which also happens at boundary layers. Here the parameter carrying the effect of the magnetic field explicitly appears in the updating scheme. Charged particles under the effect of Lorentz force, therefore, sense a reflecting boundary in a different way than uncharged particles do. This became evident when we presented the time-evolution of a charged together with an uncharged species in this Chapter. As it was a driving motivation for this thesis, we showed that the uncharged species picks up the otherwise unique behaviour of the charged species solely due to hard-core collisions, a rotational probability flux.

Self-Diffusion

4

In the preceding Chapters, we generalized the formalism of Bruna and Chapman [10–12] which included hard-core collisions into the probabilistic description of diffusion, to a generalized form of tensorial diffusion. This arises, for example, when the hard-core interacting particles are charged and submitted to a Lorentz force. The physical role of the rotational fluxes in these systems will become apparent in this Chapter[1], where we focus on the collective and self-diffusion coefficients. As the main result of this thesis we show that in contrast to ordinary systems, where the self-diffusion is reduced by collisions with other Brownian particles, collisions with the effect of a magnetic field rather can enhance the self-diffusion. We present a mechanism for this counterintuitive result and validate the theoretical predictions by Brownian dynamic simulations for dilute systems [22].

4.1 Diffusion Coefficients

There are several types of diffusion coefficients, which have to be distinguished. The collective diffusion coefficient concerns the macroscopic phenomenon of many Brownian particles diffusing simultaneously. The self-diffusion coefficient, in contrast, describes the diffusion of a single tagged particle interacting with surrounding Brownian particles [19]. There also exists the cross-diffusion coefficient, which describes the diffusion of one species of Brownian particles in presence of another. For an illustration of collective versus self-diffusion, see Figure 4.1.

The self-diffusion of a Brownian particle is generally reduced by collisions with other particles. Intuitively this can be understood as a hindrance to the random walk

---

[1] Parts of this Chapter are edited and reprinted with permission from [1]. Copyright (2022) by the American Physical Society.

E. Kalz, *Diffusion under the Effect of Lorentz Force*, BestMasters, https://doi.org/10.1007/978-3-658-39518-6_4

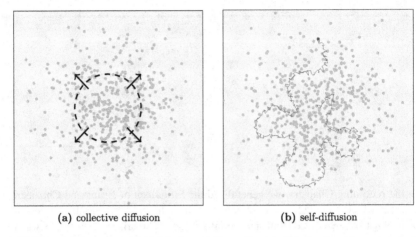

(a) collective diffusion   (b) self-diffusion

Figure 4.1 *Collective and self-diffusion.* (a) The collective diffusion is a phenomenon of many Brownian particle diffusing simultaneously. When they are Gaussian distributed, as in the schematic, collective diffusion can be associated with the growth of the Gaussian variance in time. (b) The self-diffusion, in contrast, describes the diffusion of a single Brownian particle interacting with surrounding Brownian particles. Interactions between the particles speed up the collective diffusion of the species, but hinder the self-diffusion of the tagged particle in general

of the tagged particle by other particles. The collective diffusion, in contrast, is enhanced by collisions. There exist analytically known expressions for the collective and the self-diffusion coefficients in the low-density limit [13, 20, 21].

4.1.1   Self- and Collective Diffusion

Our first focus is on obtaining the self-diffusion coefficient. Therefore we take the general approach of two diffusing species and make them identical in the presented formalism. In the two species model in Eq. (2.43), the species can differ in their Lorentz diffusivities $\frac{D_0^{(i)}}{1+\kappa_i^2}$ due to a potentially different $\kappa_i = \frac{q_i B}{\gamma_i}$. $D_0^{(i)}$ is the bare diffusivity, q_i the charge and γ_i the respective friction coefficient for each species ($i \in \{1, 2\}$). For this Chapter we set the external forces to zero.

For deriving an expression for the self-diffusion, formally we have to tag one particle in the sea of identical particles to read-off the effect of collisions. This is possible by writing the two-species model for fully identical particles, i.e. $\kappa_1 =$

$\kappa_2 \equiv \kappa$ and also $D_0^{(1)} = D_0^{(2)} \equiv D_0$. But instead of dealing with arbitrary N_1 and N_2, we assign species two to be the species of the tagged particle. Hence we set $N_1 = N$ and $N_2 = 1$, such that we have in total $N + 1$ Brownian particles in the system. The time evolution equation for p_2 now describes the time evolution of the tagged particle in presence of other identical Brownian particles, its diffusion coefficient will be the desired self-diffusion coefficient of charged particles under the effect of Lorentz force. The time evolution equation for p_1, as written from Eq. (2.41) reads

$$\frac{\partial p_1(\mathbf{x}, t)}{\partial t} = \nabla \cdot \mathsf{D} \left[\nabla p_1 + (N - 1)\sigma^2 \pi \ p_1 \nabla p_1 + \pi \sigma^2 \ (\boldsymbol{\beta} \ p_1 \nabla p_2 - \boldsymbol{\gamma} \ p_2 \nabla p_1) \right].$$
(4.1)

The time evolution equation for the tagged particle, as specified from Eq. (2.42) reads

$$\frac{\partial p_2(\mathbf{x}, t)}{\partial t} = \nabla \cdot \mathsf{D} \left[\nabla p_2 - N\pi\sigma^2 \ \boldsymbol{\gamma} \ p_1 \nabla p_2 + N\pi\sigma^2 \ \boldsymbol{\beta} \ p_2 \nabla p_1 \right].$$
(4.2)

Because of all particles are identical, the diffusion matrix multiplying both, Eqs. (4.1) and (4.2) is the same, i.e. $\mathsf{D} = \frac{D_0}{1+\kappa^2} \left(\begin{smallmatrix} 1 & \kappa \\ -\kappa & 1 \end{smallmatrix} \right)$. The matrices capturing the inter-species collisions also are the same in the two equations

$$\beta = 1 + \frac{\mathsf{D}^2}{2 \det(\mathsf{D})}, \qquad\qquad \gamma = \frac{\mathsf{D}^2}{2 \det(\mathsf{D})}.$$
(4.3)

Written in terms of the constant $\phi \equiv \frac{N\sigma^2\pi}{4} p_1$ of the remaining (identical) particles, where $\frac{N\sigma^2\pi}{4}$ is the area fraction, the tagged-particle diffusion equation (4.2) can be written as

$$\frac{\partial p_2(\mathbf{x}, t)}{\partial t} = \nabla \cdot \mathsf{D} \left[(1 - 4\phi \ \boldsymbol{\gamma}) \nabla p_2(\mathbf{x}, t) \right].$$
(4.4)

From here, the self-diffusion coefficient D_s can be read-off from the diagonal elements of the appearing diffusion tensor $\mathsf{D} (1 - 4\phi\gamma) = \mathsf{D} - 2\phi \frac{\mathsf{D}^3}{\det(\mathsf{D})}$, to be

$$D_s = \frac{D_0}{1 + \kappa^2} \left(1 - 2\phi \frac{1 - 3\kappa^2}{1 + \kappa^2} \right).$$
(4.5)

The remarkable prediction of this self-diffusion coefficient is the existence of a critical magnetic field parameter $\kappa_{\mathrm{crit}} = 1/\sqrt{3}$, above which the self-diffusion gets enhanced concerning the ideal coefficient instead of being reduced by collisions.

Table 4.1 *Mathematical structure of the self-diffusion coefficient.* If the essential matrix $\gamma = \frac{DD}{2\det(D)}$ would have had a different composition, the self-diffusion would not have a sign-changing factor. Correctly, the self-diffusion coefficient in terms of its matrix representation is given as $D_s = \text{diag}\,(D - 4\phi\,D\gamma)$, which leads to the prediction of a crossover from reduction to enhancement

γ	$D_s / \frac{D_0}{1+\kappa^2}$
$\propto DD$	$1 - 2\phi \frac{1-3\kappa^2}{1+\kappa^2}$
$\propto DD^T$	$1 - 2\phi$
$\propto D^TD$	$1 - 2\phi$
$\propto D^TD^T$	$1 - 2\phi$

Setting $\kappa = 0$ and therefore reducing the Lorentz diffusivity to the bare diffusivity D_0, we reduce the model to an uncharged system of hard-spheres. The self-diffusion reduces to the known expression $D_s = D_0\,(1 - 2\phi)$ at first-order correction in the area fraction ϕ, ignoring hydrodynamic contributions [13, 19, 20].

Let us comment at this point on one mathematical subtlety of this expression for self-diffusion. As Table 4.1 shows, the appearance of the sign-changing term $\frac{1-3\kappa^2}{1+\kappa^2}$ in the collision correction can be directly related to the cubic power of the diffusion tensor, or better say the quadratic power of the tensor in γ. In Section 2.5 we argued for which type of diffusion matrix shall multiply the first-order correction terms in the diffusion equation, and thus form the matrix γ, the tensor D itself, or its transposed D^T. Without an external drift in the system, the formalism was unable to answer this question, a result of the pure curling nature of the Lorentz-flux. Physical arguments of the impossibility of magnetic field to show a fingerprint in equilibrium gave the answer, the reason for which we included the drift-term into the presented derivation. Table 4.1 shows that the sign-changing factor in the collision correction would be absent if our conclusion would have been different.

Since the tagged particle is identical to all the other particles, we can go further by untaging the particle in focus. Mathematically we set $p_1 = p_2 \equiv p$, which formally reduces Eq. (2.43) to an one-species model. Its concentration depended diffusion coefficient now is the collective diffusion coefficient of a charged species under the effect of Lorentz force. Because of $\beta - \gamma = 1$ for this system, both, Eqs. (4.1) and (4.2) reduce to the same time evolution equation for p

$$\frac{\partial p(\mathbf{x}, t)}{\partial t} = \nabla \cdot D \left(\nabla p(\mathbf{x}, t) + N\pi\sigma^2\, p(\mathbf{x}, t)\nabla p(\mathbf{x}, t) \right). \qquad (4.6)$$

This is essentially the diffusion equation for a single species, as we have obtained it in Section 2.5, compare Eq. (2.39) with in total $N + 1$ particles.

The collective diffusion coefficient can be read-off from Eq. (4.6) to be

$$D_c = \frac{D_0}{1 + \kappa^2} \left(1 + 4\phi\right), \tag{4.7}$$

with use of the area fraction $\phi = \frac{N\sigma^2\pi}{4}p$. Again considering $\kappa = 0$ and thus reduce the Lorentz diffusivity to the bare diffusivity D_0, we obtain an uncharged system of hard-spheres in two dimensions. The collective diffusion reduces to the known expression $D_c = D_0 \left(1 + 4\phi\right)$, ignoring hydrodynamic contributions [19].

4.1.2 Two-Species Diffusion

The last class of systems, to which we want to draw attention, are systems with truly two species of particles. Here we ask the question, of how the magnetic field changes the diffusion coefficient of a tagged particle. First, we consider an uncharged particle in a sea of charged Brownian particles, which therefore experience the effect of Lorentz force. Second, we invert the system and model a charged Brownian particle in a sea of uncharged particles. These two systems seem like symmetric counterparts, which they are indeed on the two-body collision level. But due to the diffusion constant of the tagged particle being a statistical measure of all occurring collisions with the particle in focus, the diffusion constants will not show the symmetry, the systems might suggest.

The terminology, in this case, is rather unclear in comparison to the first scenarios. Clearly, these diffusion constants are not self-diffusion constants, we are dealing with two distinct species. But also the terminology of cross-diffusion does not apply, it is commonly associated with the diffusion of a whole species in presence of another species to capture the many-particle effects. In terms of the general two-species model of Eq. (2.43), the cross-diffusion corresponds to the off-diagonal terms in the matrix-systems diffusion matrix in Eq. (2.44). But this is not the route, which we take to derive the desired coefficients, instead, we follow the same approach as for deriving the self-diffusion coefficient of Eq. (4.5) in this Section. Therefore we simply tend to term them as the diffusion coefficient of the tagged particle.

We are dealing with systems of particles with identical Lorentz diffusivity $\frac{D_0^{(1)}}{1+\kappa_1^2} = \frac{D_0^{(2)}}{1+\kappa_2^2}$, but potentially different charges $\frac{q_1 B}{\gamma_1} = \kappa_1 \neq \kappa_2 = \frac{q_2 B}{\gamma_2}$. As a result, the bare diffusivities $D_0^{(1)}$ and $D_0^{(2)}$ have to account for the different charges for allowing equal Lorentz diffusivities for the two species. Besides this, we again tag one particle, representing the different species, i.e. $N_1 = N$, $N_2 = 1$, such that in total $N + 1$ particles are in the system. The time evolution equations, similar to Eqs. (4.1) and (4.2) read, for p_1

$$\frac{\partial p_1(\mathbf{x}, t)}{\partial t} = \nabla \cdot \mathsf{D}_1 \left[\nabla p_1 + (N - 1)\sigma^2 \pi \; p_1 \nabla p_1 + \pi \sigma^2 \left(\beta_1 \; p_1 \nabla p_2 - \gamma_1 \; p_2 \nabla p_1 \right) \right],$$
(4.8)

and for the tagged particles density p_2

$$\frac{\partial p_2(\mathbf{x}, t)}{\partial t} = \nabla \cdot \mathsf{D}_2 \left[\nabla p_2 - N\pi\sigma^2 \; \gamma_2 \; p_1 \nabla p_2 + N\pi\sigma^2 \; \beta_2 \; p_2 \nabla p_1 \right]. \qquad (4.9)$$

The diffusion matrices for the two species read $\mathsf{D}_i = \frac{D_0^{(i)}}{1+\kappa_i^2} \begin{pmatrix} 1 & \kappa_i \\ -\kappa_i & 1 \end{pmatrix}$ $(i \in \{1, 2\})$, which build up the matrices capturing the inter-species interactions ($i = 1, j = 2$ and vice versa)

$$\beta_i = 1 + \frac{\mathsf{D}_1 + \mathsf{D}_2}{\det(\mathsf{D}_1 + \mathsf{D}_2)} \mathsf{D}_j, \qquad \gamma_i = \frac{\mathsf{D}_1 + \mathsf{D}_2}{\det(\mathsf{D}_1 + \mathsf{D}_2)} \mathsf{D}_i. \qquad (4.10)$$

1. *charged among uncharged.* In this system, the only particle carrying a charge, is the particle in focus. Hence $\kappa_1 = 0$ and $\kappa_2 = \kappa$. The diffusivity, the tagged particle experiences can be read-off from Eq. (4.9). In terms of the area fraction of the surrounding Brownian particles $\phi = \frac{N\sigma^2 \pi}{4} p_1$ it is given as

$$D = \frac{D_0}{1 + \kappa^2} \left(1 - 8\phi \frac{1 - 2\kappa^2}{4 + \kappa^2} \right). \qquad (4.11)$$

Again we observe a critical magnetic field $\kappa_{\text{crit}} = 1/\sqrt{2}$, for which the diffusion constant of the charged particle is enhanced in comparison to its ideal diffusivity.

2. *uncharged among charged.* The particle in focus is the uncharged particle. On the two-body level this is the symmetric counterpart of the previous situation. We have $\kappa_1 = \kappa$ and $\kappa_2 = 0$. The diffusivity of the tagged particle, in terms of the area fraction of the surrounding Brownian particles ϕ is given as

$$D = D_0 \left(1 - 8\phi \frac{1}{4 + \kappa^2} \right), \qquad (4.12)$$

Table 4.2 *Diffusion coefficients of different systems.* The self-diffusion coefficient is the diffusion coefficient intrinsically attached to a tagged particle (cross). The collective diffusion coefficient represents the diffusive behaviour of many Brownian particles (no cross) simultaneously. (1a) and (1b) represent known results for uncharged hard-disks (empty circles), which the theory can reproduce. (2a) and (2b) summarize the results for the self-diffusion and collective diffusion in systems of charged hard-disks (filled circles) submitted to a Lorentz force. In row (3) and (4) the diffusion coefficients of a charged particle surrounded by an uncharged sea and vice versa are given. Interestingly the diffusion coefficients in the last two systems are not symmetric counterparts of each other

	system	sketch	diffusion constant
(1a)	self-diffusion $\kappa_1 = \kappa_2 = 0$		$D_s = D_0 \left(1 - 2\phi\right)$
(1b)	collective diffusion $\kappa_1 = \kappa_2 = 0$		$D_c = D_0 \left(1 + 4\phi\right)$
(2a)	self-diffusion $\kappa_1 = \kappa_2 \equiv \kappa$		$D_s = \frac{D_0}{1+\kappa^2} \left(1 - 2\phi \frac{1-3\kappa^2}{1+\kappa^2}\right)$
(2b)	collective diffusion $\kappa_1 = \kappa_2 \equiv \kappa$		$D_c = \frac{D_0}{1+\kappa^2} \left(1 + 4\phi\right)$
(3)	charged among uncharged $\kappa_1 \equiv \kappa, \kappa_2 = 0$		$D = \frac{D_0}{1+\kappa^2} \left(1 - 8\phi \frac{1-2\kappa^2}{4+\kappa^2}\right)$
(4)	uncharged among charged $\kappa_1 = 0, \kappa_2 \equiv \kappa$		$D = D_0 \left(1 - 8\phi \frac{1}{4+\kappa^2}\right)$

which is clearly not the symmetric analogue of the previous scenario. No enhancement is possible.

Table 4.2 summarizes the results of this chapter.

4.2 Mechanism of Enhanced Self-Diffusion

In the preceding Section, we found the self-diffusion coefficient of a charged Brownian particle under the effect of Lorentz force to be

$$D_{\mathrm{s}} = \frac{D_0}{1 + \kappa^2} \left(1 - 2\phi \frac{1 - 3\kappa^2}{1 + \kappa^2} \right), \tag{4.13}$$

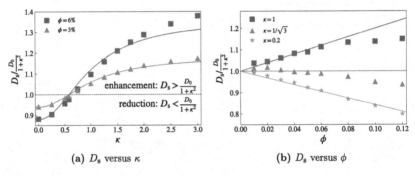

(a) D_{s} versus κ (b) D_{s} versus ϕ

Figure 4.2 *Theoretical predictions and simulation results for self-diffusion.* Reduced self-diffusion $D_{\mathrm{s}}/\frac{D_0}{1+\kappa^2}$ as a function of (a) κ and (b) area fraction ϕ, where $\frac{D_0}{1+\kappa^2}$ is the Lorentz diffusivity. (a) The self-diffusion is shown for two different area fractions ϕ, according to the many-body Brownian-dynamics simulations (symbols) [22], which were performed for $N = 500$ particles in the system. The simulation results agree with the theoretical prediction (solid lines) of a crossover regime from reduction to enhancement of the self-diffusion with a critical value at $\kappa_{\mathrm{crit}} = 1/\sqrt{3}$, where behaviour changes. (b) The self-diffusion is shown for three different values of κ. For $\kappa = 1$ ($> \kappa_{\mathrm{crit}}$) the self-diffusion increases with increasing the area fraction and for $\kappa = 0.2$ ($< \kappa_{\mathrm{crit}}$) it decreases with increasing the area fraction. For $\kappa = \kappa_{\mathrm{crit}}$ the self-diffusion remains unaffected by collisions. Simulations and theory agree on a range of $0 < \phi < 0.12$. Note that symbols cover the error bars. Figures are reprinted and edited with permission of [1]. Copyright (2022) by the American Physical Society

where $\frac{D_0}{1+\kappa^2}$ is the Lorentz diffusivity and κ the measure of Lorentz force versus drag force. Consider[1] the reduced self-diffusion $D_s/\frac{D_0}{1+\kappa^2}$. Depending on whether this is greater or smaller than one, collisions can be considered as reducing or enhancing the self-diffusion of the particle respectively. Our theory predicts that there is a crossover behaviour governed by the magnetic field, such that for small κ, collisions reduce the self-diffusion, whereas for large κ there is an enhancement. There exists a critical magnetic field, corresponding to a $\kappa_{crit} = 1/\sqrt{3}$ at which $D_s/\frac{D_0}{1+\kappa^2}$ is exactly one, implying the absence of a collision effect. The behaviour of the reduced self-diffusion $D_s/\frac{D_0}{1+\kappa^2}$ versus κ (for fixed ϕ) is shown in Figure 4.2 (a), where the predictions are veryfied by many-body Brownian-dynamics simulations [22]. There exists a different interpretation of the above result. Showing the reduced self-diffusion $D_s/\frac{D_0}{1+\kappa^2}$ versus ϕ (for fixed κ) in Figure 4.2 (b) we can identify the three regimes of magnetic fields. $\kappa < \kappa_{crit}$, where the self-diffusion reduces with increasing the density of particles, and hence more collisions in the system. $\kappa > \kappa_{crit}$, where the self-diffusion surprisingly increases with increasing the denisty of particles and the predicted crossover value of κ_{crit}, where the particles effectively become invisible to each other. Again, the predictions are veryfied by many-body Brownian-dynamics simulations [22].

In order to understand the mechanism behind the changing behaviour of self-diffusion, first we want to highlight the similarities between an overdamped and an underdamped system under the effect of Lorentz force. Whereas in an underdamped system the Lorentz force curves particle trajectories, the affected quantity in an overdamped system is the particle-flux. The off-diagonal elements in the diffusion tensor $D = \frac{D_0}{1+\kappa^2}\begin{pmatrix} 1 & \kappa \\ -\kappa & 1 \end{pmatrix}$ due to Lorentz force add a curl to the probability flux $J = D\nabla p = \frac{D_0}{1+\kappa^2}\left[\nabla p + \kappa \nabla p \times \hat{n}_B\right]$ with $\hat{n}_B$ as the normal vector of the magnetic field. It is this additional rotational Lorentz-flux [15, 17] which is the conceptual analogue to the trajectory curving in an underdamped system. The flux is divergence free, but generates a new type of boundary conditions [17]. These oblique boundary conditions are characterized by a flow along the boundary, as we have emphasized in Section 3.1.

The physical reason for the qualitative change in the self-diffusion coefficient is that the Lorentz force changes the collision statistics in the hard-core system unexpectedly. In an uncharged system with hard-core interactions ($\kappa = 0$), collisions hinder a particle in its free diffusion, the particles are effectively reflected, whenever they collide (see Figure 4.3, left column). Hence the self-diffusion is reduced in first-order contribution in area fraction, $D_s = D_0(1-2\phi)$. At small magnetic fields (small

---

[1] This section served as the basis for a distinct report and is reprinted with permission from [1]. Copyright (2022) by the American Physical Society.

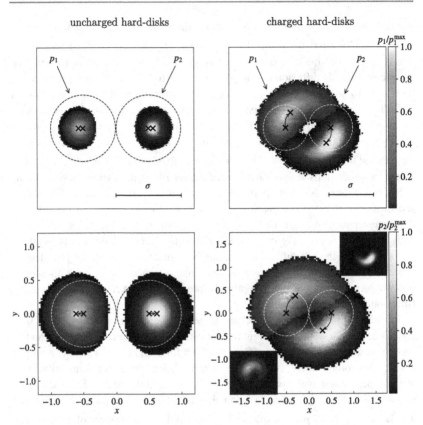

Figure 4.3 *Ying-Yang figure: collisions of hard-disks.* Scaled one-body probability densities p_1/p_1^{max} and p_2/p_2^{max} of two Brownian particles of diameter $\sigma/L = 1$, initially placed at contact distance (dashed lines). The motion of the average centre-of-mass coordinate is indicated by the crosses. Two snapshots are shown in reduced time $\tilde{t} = \frac{D_0}{(1+\kappa^2)L^2}t$. The first row corresponds to $\tilde{t}_1 = 0.02$, and the second to $\tilde{t}_2 = 0.04$. The left column represents uncharged particles ($\kappa = 0$), undergoing hard-core collisions. As it is apparent, a collision hinders each particle to explore the space in a free diffusive way. On the contrary, the right column shows charged particles ($\kappa = 5$) diffusing under the effect of Lorentz force. Here collisions between the particles result in a probability flow around the counterpart of collisions. The insets show hidden parts of each particle's curved probability distribution due to the overlap in the main figure. The presented data originate from two-body Brownian-dynamics simulations of five million realizations. Note that, when comparing the insets and the main figures, the presented data are truncated to show the two probability distributions simultaneously in one figure. Negligible probabilities are cut-off for both particles, which results in a white background. Figures are reprinted and edited with permission of [1]. Copyright (2022) by the American Physical Society

κ) still, the radial backward flux from the collision place is dominating. But with increasing κ, the rotational Lorentz-flux becomes more prominent, and suddenly favours to curve around the collision partner (see figure 4.3, right column). As it can be seen in the insets of Figure 4.3, there is a macroscopic probability for the two particles to interchange positions due to the collision.

Take for example the x-component of the probability flux $\mathbf{J}$. In case of an unperturbed diffusion equation, $J_x = \partial_x p + \kappa \partial_y p$. κ is the parameter controlling the strength of the coupling between x- and y-derivatives of the probability density p. It is apparent that with increasing κ the contribution of the y-derivative will overtake the contribution of the x-derivative to form J_x. A movement in x-direction results in a stronger movement in y-direction after a collision for a high enough κ. In the bulk, these antisymmetric contributions cancel each other in the diffusive process. But each collision-partner can be seen as a (moving) boundary, for which we have shown in Chapter 3 that antisymmetric effects do not cancel. Therefore, when the magnetic field becomes high enough, eventually more than one collision is induced. The process repeats until the particles have interchanged, due to them under permanent collision rolling around each other. By this, we can understand, why the self-diffusion gets enhanced in comparison to the ideal diffusivity instead of only approaching it asymptotically with increasing κ.

Out of the perspective of one particle, it can enhance its self-diffusion, since a collision now supports exploring the space behind the colliding counterpart, instead of hindering it. Out of the statistical perspective of the whole species, nothing has changed, except for two particles having interchanged positions due to a collision. Hence, beyond a rescaling of the diffusion time-scale, the collective diffusion remains unaffected. This is in agreement with the prediction of the theory, $D_c / \frac{D_0}{1+\kappa^2} = 1 + 4\phi$.

Instead of only considering all identical particles, the theory allows for particles to be of a different type. Specifically, we obtained the diffusion coefficient of a tagged particle in presence of a different species.

We showed that the individual diffusion coefficient of an uncharged Brownian particle diffusing in a sea of particles submitted to a Lorentz force is given as $D = \frac{D_0}{1+\kappa^2} \left(1 - 8\phi \frac{1}{4+\kappa^2} \right)$. As case (b) in Figure 4.4 also illustrates, the uncharged particle never can enhance its diffusion, only asymptotically approach the ideal diffusivity. The tagged particle cannot benefit from the collisions in the way, charged particles do. This scenario is in full agreement with the suggested mechanism.

We also modelled the case of a charged particle diffusing in a sea of ordinary, uncharged Brownian particles, for which $D = \frac{D_0}{1+\kappa^2} \left(1 - 8\phi \frac{1-2\kappa^2}{4+\kappa^2} \right)$ shows a crossover from reduction to enhancement at $\kappa_{\text{crit}} = 1/\sqrt{2}$, see also Figure 4.4 (c).

Here, in contrast, the charged particle can benefit from the collisions under a high enough magnetic field, and hence show the crossover to enhanced self-diffusion concerning its ideal diffusivity. The latter two systems, even though being the symmetric counterparts in a two-body collisions-picture emphasize the relevance of the particle in focus.

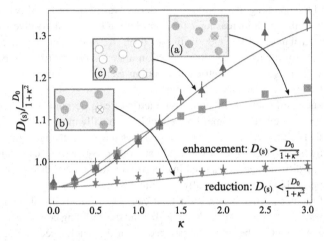

Figure 4.4 *(Self-) Diffusion for different scenarios.* The dimensionless (self-) diffusion coefficient $D_{(s)}$ for a constant area fraction ($\phi = 3\%$) of hard-disks submitted to a Lorentz force. In the sketches, the empty disks represent uncharged Brownian particles ($\kappa = 0$), and the filled disks are charged particles with a finite κ. The tagged particle carries a cross. Shown are the three different scenarios of (a) charged disks, where the self-diffusion changes behaviour from reduction to enhancement due to two-body collisions at a critical value of $\kappa_{crit} = 1/\sqrt{3}$. (b) an uncharged particle in a sea of charged particles of identical bare diffusivity, where the diffusion coefficient of the tagged particle only is reduced in comparison to the ideal (non-interacting) case, but asymptotically approaches that with increasing κ. And (c), a charged particle in a sea of uncharged particles with identical bare diffusivity. Again there exists a critical value of $\kappa_{crit} = 1/\sqrt{2}$ where the diffusion coefficient of the tagged particle changes behaviour from reduction to enhancement due to collisions with increasing κ. The simulation results (symbols) [22] agree with the theoretical predictions (lines). Note that for the squares symbols cover the error bars. The Figure is reprinted and edited with permission of [1]. Copyright (2022) by the American Physical Society

4.3   Summary

In this Chapter, we presented the main result of this work. By specifying the two-species theory of diffusing particles with hard-core interactions, we derived time-evolution equations for the one-body densities of all-identical particles. The formalism allowed us to tag one particle, of which diffusion equation we could read-off the collision corrections to the self-diffusion coefficient under Lorentz force. By furthermore untagging this particle we obtained the collective diffusion coefficient. The derived self-diffusion predicted a crossover regime from reduction to enhancement, governed by collisions and the magnetic field. We presented the results of Brownian-dynamics simulations, which validated our predictions. In line with the presented mechanism of this crossover, we also derived the diffusion coefficient of an uncharged particle in a sea of charged particles and vice versa. All presented systems, including the predicted collective diffusion for the all-identical charged species, support the presented mechanism. Charged particles benefit from collisions since the counterpart in a collision can be viewed as a (moving) boundary. From the previous Chapter 3 we know that when a charged particle under Lorentz force diffuses in presence of a boundary, the off-diagonal contributions from the diffusion tensor due to Lorentz-force do not cancel, but instead couple x- and y-direction. It is essentially this coupling, which lets charged particles benefit from a collision, instead of getting hindered by it, as this is the case for ordinary colloidal systems.

First-Principles Approach to Self-Diffusion   5

The prediction of a self-diffusion-coefficient, which changes sign in the first-order correction in area fraction with increasing the magnetic field strength κ was a very surprising and unexpected result of this work. Validating this by measuring the mean-squared displacement in a dilute system of charged particles under Lorentz force in simulations was a necessary next step [22]. But since neither the asymptotic model nor the simulation results in the first attempts gave a clear hint to the underlying physical principles of this counter-intuitive result, we undertook another theoretical approach, completely different from the presented asymptotic theory. It is not the aim of this thesis, nor this Chapter[1] to present this theory in detail. Its often purely mathematical character and simply the necessary length of a potential presentation led to the decision to not include any details of this first-principles approach here. Nevertheless, since the results of this approach are interesting by themselves, we give a brief survey on the calculation and its main results.

Following Hanna, Hess and Klein [20, 30], we can write the time evolution equation for the joint probability distribution $P(t) \equiv P(\mathbf{r}_1, \mathbf{r}_2, t)$ of two Brownian particles at position $\mathbf{r}_1$ and $\mathbf{r}_2$ as

$$\frac{\partial}{\partial t} P(t) = \nabla_1 \cdot \mathsf{D} \left[\nabla_1 + \beta \, \nabla_1 U(\mathbf{r}_1, \mathbf{r}_2) \right] P(t) + \nabla_2 \cdot \mathsf{D} \left[\nabla_2 + \beta \, \nabla_2 U(\mathbf{r}_1, \mathbf{r}_2) \right] P(t),$$

$$(5.1)$$

where ∇_i denotes the partial differentiation with respect to particle coordinate $\mathbf{r}_i$ for $i \in \{1, 2\}$ and β is the inverse temperature. The particles are equal sized with diameter σ and interact with each other via a hard-core potential $U(\mathbf{r}_1, \mathbf{r}_2)$. We are

---

[1] Parts of this Chapter are edited and reprinted with permission from [1]. Copyright (2022) by the American Physical Society.

© The Author(s), under exclusive license to Springer Fachmedien Wiesbaden GmbH,   47
part of Springer Nature 2022
E. Kalz, *Diffusion under the Effect of Lorentz Force*, BestMasters,
https://doi.org/10.1007/978-3-658-39518-6_5

interested in the diffusive behaviour of these particles when they are charged. Due to the present Lorentz force, which is induced by a magnetic field perpendicular to the two-dimensional plane of diffusion we are in the regime of tensorial diffusion $D = \frac{D_0}{1+\kappa^2} \left(\begin{smallmatrix} 1 & \kappa \\ -\kappa & 1 \end{smallmatrix} \right)$. The diffusion tensor describes an isotropic space with Lorentz diffusivity $\frac{D_0}{1+\kappa^2}$, whereas the effect of the Lorentz force sits in the anti-symmetric off-diagonal elements κ.

Because of the isotropy in the system, we can decompose the equation governing the time evolution of the joint probability of the two particles into an equation governing the probability distribution of the center of mass coordinate $\mathbf{R} = (\mathbf{r}_1 + \mathbf{r}_2)/2$ and the inner coordinate $\mathbf{r} = (\mathbf{r}_1 - \mathbf{r}_2)$. Eq.(5.1) decouples into

$$\frac{\partial}{\partial t}\rho(\mathbf{R}, t) = \frac{1}{2}\nabla_\mathbf{R} \cdot \left[\frac{D_0}{1 + \kappa^2} \nabla_\mathbf{R} \right] \rho(\mathbf{R}, t), \tag{5.2}$$

$$\frac{\partial}{\partial t}\varrho(\mathbf{r}, t) = 2\nabla_\mathbf{r} \cdot \left[\frac{D_0}{1 + \kappa^2} \mathbb{1}\nabla_\mathbf{r} + \beta \, D\nabla_\mathbf{r}U(\mathbf{r}) \right] \varrho(\mathbf{r}, t), \tag{5.3}$$

where $\nabla_\mathbf{R}$, $\nabla_\mathbf{r}$ denote partial derivatives with respect to the coordinates $\mathbf{R}$, $\mathbf{r}$. $\rho(\mathbf{R}, t)$ and $\varrho(\mathbf{r}, t)$ denote the distributions of the center of mass coordinate and the inner coordinate, respectively.

It appears that the nature of the hard-core collisions as well as the modification due to the opposite off-diagonal elements in the diffusion tensor only affect the inner coordinate. The equation for the centre of mass coordinate becomes rather trivial and admits a Gaussian solution with a scalar diffusion coefficient $\frac{1}{2}\frac{D_0}{1+\kappa^2}$. Furthermore, the hard-core interaction potential takes up a simple form in the inner coordinate $\mathbf{r} = r\hat{\mathbf{r}}$ as $U(r) = \left\{ \begin{smallmatrix} 0, & r>\sigma \\ \infty, & r\leq\sigma \end{smallmatrix} \right.$. Mathematically this singular interaction potential is implemented cleverly in terms of a Heaviside step-function $\Theta(r - \sigma)$, i.e. by never allowing the inner coordinate to overlap with the particle diameter σ. Because of this extraction of the step-function, the derivation allows little physical insight. The objects of interest become rather mathematical.

We are interested in solving for the conditional distribution of the inner coordinate $\varrho(\mathbf{r}, t|\mathbf{r}_0, t_0)$, which is the one-body probability distribution of finding particle two at time t at a distance $\mathbf{r}$ away from particle one, given that initially at $t_0 = 0$ they were at a distance $|\mathbf{r}_0| > \sigma$. Note that the derived problem for the inner coordinate of two particles in the presence of Lorentz force naturally produce the oblique no-flux boundary condition, one of the properties unique to the diffusion under Lorentz force (see also Eq.(3.1)). Including the boundary condition, the problem can be solved exactly and the solution is given in the Laplace-domain as

$$
\tilde{\varrho}(\mathbf{r}, s|\mathbf{r}_0) = \Theta(r - \sigma) \sum_{n=-\infty}^{\infty} \frac{e^{in(\varphi - \varphi_0)}}{2\pi} \frac{\sigma^2}{2D_0/(1+\kappa^2)} \int_0^\infty du \, u \frac{J_n(ur_0)}{z^2 + u^2\sigma^2}
$$

$$
\times \left[J_n(ur) - K_n\left(\frac{zr}{\sigma}\right) \frac{u\sigma \, J_n'(u\sigma) + in\kappa \, J_n(u\sigma)}{z \, K_n'(z) + in\kappa \, K_n(z)} \right]. \tag{5.4}
$$

Here i denotes the imaginary unit, u is a dummy integration variable and we introduced z for a shortness of notation as a modified Laplace-variable $z = \sqrt{\frac{s\sigma^2}{2D_0/(1+\kappa^2)}}$, s denotes the original Laplace variable. The inner coordinate is decomposed into its polar representation in two dimensions according to $\mathbf{r} = (r, \varphi)$, similarly $\mathbf{r}_0 = (r_0, \varphi_0)$. $J_n(x)$ is the Bessel function of first kind and $K_n(x)$ the modified Bessel function of third kind, both of order n. We have used the notation $f'(c) = \frac{df(x)}{dx}\big|_{x=c}$ as a shorthand notation of a derivative of a function f evaluated at a value c.

Only considering up to two-body collisions, we can derive a Green-Kubo like relation for the self-diffusion D_s with the time-integral of a correlation function. For uncharged systems, i.e. $\kappa = 0$ and hence a scalar diffusion constant $\mathsf{D} = D_0\mathbf{1}$, the force autocorrelation (FAC) function $C_F^{\kappa=0}(t) = \frac{1}{2}\langle \mathbf{F}(0) \cdot \mathbf{F}(t) \rangle$ as an ensemble average of force $\mathbf{F} = -\nabla_\mathbf{r} U(r)$ at time t projected on the force at initial time $t = 0$ appears as the integrand. In contrast, for the system as described by the anti-symmetric diffusion tensor, full information of the FAC tensor is needed to write the same equation. It appears that the tensor with anti-symmetric off-diagonal elements has a similar structure as the diffusion tensor. The relation between self-diffusion and FAC tensor reads

$$
D_s = \frac{D_0}{1+\kappa^2} \left(1 - \frac{\beta^2}{2D_0^2}(\mathsf{D}^2)^\mathsf{T} : \int_0^\infty dt \begin{pmatrix} C_F^{sym}(t) & C_F^{asym}(t) \\ -C_F^{asym}(t) & C_F^{sym}(t) \end{pmatrix} \right) \tag{5.5}
$$

$$
= \frac{D_0}{1+\kappa^2} \left(1 - \beta^2 \int_0^\infty dt \, (1 - \kappa^2) \, C_F^{sym}(t) - 2\kappa \, C_F^{asym}(t) \right). \tag{5.6}
$$

Here $\mathbf{A} : \mathbf{B}$ denotes the double-contraction between two tensors $\mathbf{A}$ and $\mathbf{B}$. This was employed when going from line (5.5) to (5.6).

The integrand $C_F(t) = (1 - \kappa^2) \, C_F^{sym}(t) - 2\kappa \, C_F^{asym}(t)$ reduces for an uncharged system ($\kappa = 0$) to $C_F^{sym}(t)$, which equals the ordinary FAC function $C_F^{\kappa=0}(t)$.

Therefore we call $C_F(t)$ a modified force autocorrelation (mFAC) function. To solve the integral, we change into the Laplace-domain, where a time integral from zero to infinite corresponds to a limit of the Laplace variable s going to zero. The contributions to the mFAC function from the symmetric and anti-symmetric parts of the FAC tensor can be calculated analytically, and read in Laplace-domain

$$\tilde{C}_F^{\text{sym}}(s) = \frac{1}{\beta^2} \frac{N}{V} \int d\mathbf{r} \int \mathbf{r}_0 \, \delta(r - \sigma)\delta(r_0 - \sigma) \, \tilde{\varrho}(\mathbf{r}, s|\mathbf{r}_0) \, \hat{\mathbf{r}} \cdot \mathbb{1}\hat{\mathbf{r}}_0, \qquad (5.7)$$

$$\tilde{C}_F^{\text{asym}}(s) = \frac{1}{\beta^2} \frac{N}{V} \int d\mathbf{r} \int \mathbf{r}_0 \, \delta(r - \sigma)\delta(r_0 - \sigma) \, \tilde{\varrho}(\mathbf{r}, s|\mathbf{r}_0) \, \hat{\mathbf{r}} \cdot \boldsymbol{\varepsilon}\hat{\mathbf{r}}_0, \qquad (5.8)$$

where the conditional distribution of the inner coordinate $\tilde{\varrho}$ of Eq.(5.4) appears explicitly. $\tilde{C}_F^{\text{sym}}$, as the symmetric part of the FAC tensor, gets contribution from the symmetric part $\hat{\mathbf{r}} \cdot \mathbb{1}\hat{\mathbf{r}}_0$ of the dyadic tensor $\hat{\mathbf{r}} \otimes \hat{\mathbf{r}}_0$, formed by the outer product of the unit vectors $\hat{\mathbf{r}}$ and $\hat{\mathbf{r}}_0$. Here $\mathbb{1} = \left(\begin{smallmatrix} 1 & 0 \\ 0 & 1 \end{smallmatrix}\right)$ is the identity tensor. The anti-symmetric part of the FAC tensor, $\tilde{C}_F^{\text{asym}}$, in contrast, captures the antisymmetric part of the outer product $\hat{\mathbf{r}} \cdot \boldsymbol{\varepsilon}\hat{\mathbf{r}}_0$, where $\boldsymbol{\varepsilon} = \left(\begin{smallmatrix} 0 & 1 \\ -1 & 0 \end{smallmatrix}\right)$ is the fully anti-symmetric Levi-Chivita symbol in two dimensions. This also can be interpreted physically in the following sense; the anti-symmetric part of the dyadic tensor equals a cross-product of the unit-vectors $\hat{\mathbf{z}} \cdot (\hat{\mathbf{r}} \times \hat{\mathbf{r}}_0)$, which, strictly speaking is undefined in two dimensions. Therefore it is projected into the plane of diffusion by the magnetic field direction, which is taken to point along the auxiliary third $\hat{\mathbf{z}}$-direction, i.e. $\mathbf{B} = B\hat{\mathbf{z}}$. Hence, $\tilde{C}_F^{\text{asym}}$ can be interpreted as the shadow of the additional Lorentz force in the system. Note that the symmetric and anti-symmetric part of the FAC tensor only contain contributions from the radial components when being equal to the particle diameter σ. The appearing Dirac delta functions $\delta(r - \sigma)$ and $\delta(r_0 - \sigma)$ ensure that the singular interaction potential of hard particles only contributes at contact distance $r = \sigma, r_0 = \sigma$ to the mFAC function.

Using the form of the inner conditional probability density $\tilde{\varrho}$, we can evaluate the symmetric and anti-symmetric part of the FAC tensor in Eqs.(5.7) and (5.8) and express them in terms of modified Bessel functions $K_n(z)$ of order $n = 0, 1$

$$\tilde{C}_F^{\text{sym}}(s) = \frac{2\phi}{\beta^2 \frac{D_0}{1+\kappa^2}} \frac{K_1(z) \, (zK_0(z) + K_1(z))}{(zK_0(z) + K_1(z))^2 + (\kappa \, K_1(z))^2}, \qquad (5.9)$$

$$\tilde{C}_F^{\text{asym}}(s) = \frac{2\kappa\phi}{\beta^2 \frac{D_0}{1+\kappa^2}} \frac{K_1^2(z)}{(zK_0(z) + K_1(z))^2 + (\kappa \, K_1(z))^2}, \qquad (5.10)$$

where again z denotes the modified Laplace-variable $z = \sqrt{\frac{\sigma^2 s}{2D_0/(1+\kappa^2)}}$ and $\phi = \frac{N}{V}\frac{\pi\sigma^2}{4}$ is the area fraction in two dimensions. The functions $\tilde{C}_F^{\text{sym}}$ and $\tilde{C}_F^{\text{asym}}$ reduced by the factor $(\beta^2 \frac{D_0}{1+\kappa^2})/\phi$ are numerically inverted and plotted as functions in real time t in Figure 5.1 (a) and (b) for different values of κ. In Figure 5.1 (c) we plot the whole mFAC function C_F as formed by C_F^{sym} and C_F^{asym} according to Eq.(5.6).

The self-diffusion is analytically accessible by taking the Laplace-variable $s \to 0$ limit in Eq.(5.6) and making use of the evaluated forms of $\tilde{C}_F^{\text{sym}}$ and $\tilde{C}_F^{\text{asym}}$. It turns out that the self-diffusion is in complete agreement with the predictions of the asymptotic model

$$D_s = \frac{D_0}{1+\kappa^2}\left(1 - 2\phi\frac{1-3\kappa^2}{1+\kappa^2}\right). \tag{5.11}$$

When we now remember, that C_F^{sym} coincides with the ordinary FAC function for uncharged systems, it is interesting to note that with increasing κ, C_F^{sym} turns negative in time, as it can be seen from Figure 5.1 (a). The FAC function turning negative in time originates from the mutual rolling of particles around each other. Increasing κ, the negative-turning becomes more pronounced; the direction of the force that particles experience due to collisions reverses and these negative correlations persist over increasing time and increasing κ. Hence the FAC function gave birth to the idea of particles effectively rolling around due to collisions and this finally completed the picture as it is presented in this thesis.

Furthermore, the finite behaviour of C_F^{asym} in $t \to 0$ limit, as apparent from Figure 5.1 (b), is an unexpected behaviour for an autocorrelation function, and we still have not found an explanation for this. Also C_F shows a fingerprint of this unusual behaviour. The function undergoes a change in its $t \to 0$ behaviour when crossing $\kappa = 1$ from a function behaving like a usual autocorrelation function for hard-core interacting particles, i.e. with positive divergence in the zero-time limit ($\kappa < 1$), to a function with finite value at zero-time ($\kappa = 1$) and up to a function with negative divergence in zero-time limit ($\kappa > 1$). It is apparent from Eq.(5.6) that this behaviour is induced by the interplay of the $(1 - \kappa^2)$ term, originating in the squared diffusion tensor, together with the finite values in zero-time limit of C_F^{asym}. We have plotted C_F for different values of κ in Figure 5.1 (c), where the change in behaviour around $\kappa = 1$ is visible.

Besides the as yet missing understanding of these interesting predictions of the mFAC function for hard particles, we also would be interested in finding an exact analytical inversion [31] into real-time of the expressions for $\tilde{C}_F^{\text{sym}}$ and $\tilde{C}_F^{\text{asym}}$. By this, we hope to find the short- and long-time behaviour or even intermediate time scales in the decay of the mFAC function. To our knowledge, even in the uncharged

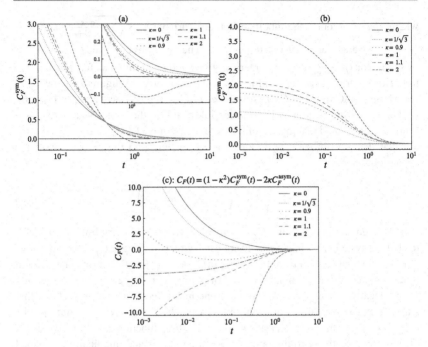

Figure 5.1 *Modified force autocorrelation function for hard-disks.* Exact calculations provide an analytic result for the modified force autocorrelation (mFAC) function $C_F(t)$ in the two particle system of hard-disks under the effect of Lorentz force. (c) With increasing strength of the magnetic field (increasing the parameter κ) the mFAC function changes behaviour from a purely positively decreasing function (for $\kappa = 0$) in reduced time t/τ_s, where $\tau_s = 2D_0/(\sigma^2(1 + \kappa^2))$, to a function with zero-crossing ($\kappa < 1$). Further increasing κ, the mFAC function becomes finite at $t \rightarrow 0$ ($\kappa = 1$) and even diverges negatively for $t \rightarrow 0$ ($\kappa > 1$). This unusual behaviour of an autocorrelation function is governed by the symmetric (C_F^{sym}) and anti-symmetric (C_F^{asym}) parts of the FAC tensor, as shown in (a) and (b). C_F^{sym} turns negative with increasing κ (inset in (a)), in full agreement with the collision induced rolling effect to explain the enhanced self-diffusion and C_F^{asym} is finite in $t \rightarrow 0$, a behaviour still to be understood. All functions are shown in reduced form with the system-parameter: $(\beta^2 \frac{D_0}{1+\kappa^2})/\phi$. Figure (a) is reprinted and edited with permission of [1]. Copyright (2022) by the American Physical Society

case, the long- and short-time behaviour of the FAC function of hard-disks are not very much investigated in two dimensions.

Conclusions and Outlook

6

In this thesis, we showed that in systems under the effect of Lorentz force, which are characterized by diffusion tensors with antisymmetric elements, collisions surprisingly can enhance self-diffusion. This result is in contrast to previous analytical and simulation work, which showed that in ordinary colloidal systems the self-diffusion is reduced by collisions. In systems under Lorentz force, due to an inherent curving effect, the motion of particles is facilitated, instead of hindered by collisions. Consistent with this we found that the collective diffusion remains unaffected. Furthermore, an uncharged Brownian particle in a sea of charged particles cannot benefit from the collision in such a way and hence cannot enhance its self-diffusion. Based on a geometrical model, we theoretically predicted this effect of a magnetic field governed crossover from a reduced to an enhanced self-diffusion. The physical interpretation we presented is quantitatively supported by the force autocorrelation function, which turns negative with an increasing magnetic field. Using Brownian-dynamics simulations, we validated the predictions.

In Chapter two we shortly reviewed the intricate procedure to obtain an over-damped equation for a particle diffusing under the effect of Lorentz force. This equation and the corresponding probabilistic description itself already were the object of scientific debate. We took this frontier of research as a starting point for including interactions into the picture. By a procedure, based on geometrically treating hard-core interactions, we derived time-evolution equations for the one-body probabilistic distribution of particles of different species. These equations included the explicit contributions of inter- and intra-species collisions to the diffusive process.

In Chapter three we presented aspects of the numerical implementation of the obtained equations. Numerically solving for the diffusion of particles near a boundary required special treatment. We went on by exploring the far-reaching effect of diffusion with finite-size effects. Therefore we studied the simultaneous time evolution of a charged and an uncharged species in the same system. We showed that the uncharged species takes over the unique rotative effect the Lorentz force generates.

In Chapter four we applied the model to all-identical particles, taking advantage of the two-species nature, by effectively tagging one particle. By this, we derived the remarkable result of a self-diffusion coefficient for charged Brownian particles under the effect of Lorentz force

$$D_s = \frac{D_0}{1 + \kappa^2} \left(1 - 2\phi \frac{1 - 3\kappa^2}{1 + \kappa^2} \right), \tag{6.1}$$

to show a magnetic field governed crossover from reduction to enhancement with an increase in the strength of the magnetic field. We presented a mechanism for this counterintuitive result. This was also supported by different scenarios, which we addressed with the same formalism. We validated the predictions by Brownian-dynamics simulations.

In the final Chapter five, we summarized a different, first-principles, approach to the self-diffusion of charged particles under the effect of Lorentz force. The theory predicts the same result for the self-diffusion as the asymptotic model. Moreover, the first-principles approach also gave access to the force autocorrelation function, which arises in the collision statistics of particles. This was the crucial step in finally understanding the fundamental physical mechanism as we presented it in Chapter four.

Binary systems with two diffusing species are an active field of research. One recent work by Grosberg and Joanny dealt with particles, where each species is subjected to a different thermostat [32]. They predicted a temperature-ratio-driven phase separation. Our theory is also able to consider the two-temperature system, even without an applied magnetic field. Relating their formalism to ours, it further-more becomes evident that we have an additional term, representing an inter-species contribution to the first-order correction in area fraction. It would be interesting to investigate, whether our model can reproduce the temperature-driven phase sep-aration. With their formalism they were also able to report the known depletion separation, hard-disks undergo for a certain size ratio. Even though, we have mod-elled equal-sized particles, generalizing to different sizes is not difficult within the asymptotic theory.

Another work, closely related to the development of the asymptotic theory itself will be to generalize to three-body collisions of charged particles under the effect of Lorentz force. They are of importance when particles diffuse in narrow geome-tries, such as channels, where one dimension becomes comparable with the particle size. Furthermore, it would be of interest, how higher-order corrections change the self-diffusion. Our physical model, as such, breaks down, when three particles are colliding. Including three-body collisions would further allow predicting the dynamical behaviour of charged particles even for higher densities, and search for dynamical caging effects of uncharged particles surrounded by charged particles under Lorentz force, one of the early driving ideas in this project.

Thinking of modified diffusion coefficients, another generalization of the theory is to model short-range attractive or repulsive interaction potentials between the particles. Work done by Maria Bruna [10] already showed, that in a single species system, the generalization to soft-interaction potentials produces a numerical correction in terms of an appearing Mayer-function integral in the equations. We wonder whether similar corrections appear in a system of charged particles under the effect of Lorentz force.

When modelling the diffusion under Lorentz force in this thesis, we have restricted the analysis to the two-dimensional plane of diffusion, projected out by the direction of the magnetic field. But extending the theory to three dimensions, the space and hence the diffusion tensor becomes anisotropic due to the presence of the magnetic field

$$
\mathsf{D} = \frac{D_0}{1 + \kappa^2} \begin{pmatrix} 1 & \kappa & 0 \\ -\kappa & 1 & 0 \\ 0 & 0 & 1 + \kappa^2 \end{pmatrix}.
\tag{6.2}
$$

Already the two-body problem is not solvable with the methods as we have used them. In this scenario, the equation has a different symmetry than the boundary conditions. We wonder whether we can expect new interesting physics to arise in three dimensions.

Turning to the first-principles approach, we also see the potential for future work. As we pointed out in Chapter five, as yet, the force autocorrelation function is only given in the Laplace domain. Analytical inversion remains a task, not only of mathematical interest. The exact real-time autocorrelation function would provide us especially their short- and long-time behaviour. To our knowledge, even in the uncharged case, the asymptotic behaviour of the force autocorrelation function of hard-disks is not much investigated in two dimensions.

But also the exact inner solution itself appears in a variety of related problems; for example, when studying the first passage time of a particle. Let $q(\mathbf{r}, t | \mathbf{r}_0, t_0 = 0)$ denote the probability density function for a particle to reach the point $\mathbf{r}$ for the first time at time t, given that it started at $\mathbf{r}_0$. In the Laplace-domain, q can straigthforwardly be related to the conditional probability distribution $\tilde{\varrho}(\mathbf{r}_1, s | \mathbf{r}_0)$ of the first-principles approach [24]

$$\tilde{q}(\mathbf{r}, s|\mathbf{r}_0) = \frac{\tilde{\varrho}(\mathbf{r}_1, s|\mathbf{r}_0)}{\tilde{\varrho}(\mathbf{r}_1, s|\mathbf{r})}. \tag{6.3}$$

Here the variable $\mathbf{r}_1$ represents any position, after going from $\mathbf{r}_0$ to $\mathbf{r}$.

At the date of publishing this thesis, we have already prepared two distinct reports. One mainly covers the asymptotic theory and its results [1], whereas the other publication, which discusses the first-principles approach is being prepared right now.

It is at this point, that I want to thank for the motivating, thrilling and hours lasting discussions which I was pleased to have with my supervisor Dr. Abhinav Sharma. A lot of my results would not exist without his encouraging way of motivating his students and also helping them over non-satisfying times, as theoreticians, as I want to become one, often have. I am looking forward to many fruitful collaborations in near and far future research.

Bibliography

[1] Erik Kalz, Hidde Derk Vuijk, Iman Abdoli, Jens-Uwe Sommer, Hartmut Löwen, and Abhinav Sharma. Collisions enhance self-diffusion in odd-diffusive systems. *Phys. Rev. Lett.*, 129:090601, Aug 2022.

[2] Robert Brown. XXVII. A brief account of microscopical observations made in the months of June, July and August 1827, on the particles contained in the pollen of plants; and on the general existence of active molecules in organic and inorganic bodies. *The Philosophical Magazine*, 4(21):161–173, 1828.

[3] Robert Brown. XXIV. Additional remarks on active molecules. *The Philosophical Magazine*, 6(33):161–166, 1829.

[4] Stephen George Brush. A history of random processes. *Archive for history of exact sciences*, 5(1):1–36, 1968.

[5] Albert Einstein. Über die von der molekularkinetischen Theorie der Wärme geforderte Bewegung von in ruhenden Flüssigkeiten suspendierten Teilchen. *Annalen der Physik*, 4, 1905.

[6] Jean Perrin. Mouvement brownien et réalité moléculaire. 1909.

[7] Marian von Smoluchowski. Zur kinetischen Theorie der Brownschen Molekularbewegung und der Suspensionen. *Annalen der Physik*, 326(14), 756–780, 1906.

[8] Adriaan Daniël Fokker. *Over Brown'sche Bewegingen in het Stralingsveld, en Waarschijnlijkheids-Beschouwingen in de Stralingstheorie.* Joh. Enschedé en Zonen, 1913.

[9] Jean-Pierre Hansen and Ian Ranald McDonald. *Theory of simple liquids: with applications to soft matter.* Academic Press, 2013.

[10] Maria Bruna. *Excluded-volume effects in Stochastic Models of Diffusion.* PhD thesis, University of Oxford, St. Anne's College, 2012.

[11] Maria Bruna and Stephen Jonathan Chapman. Excluded-volume effects in the diffusion of hard spheres. *Physical Review E*, 85(1):011103, 2012.

[12] Maria Bruna and Stephen Jonathan Chapman. Diffusion of multiple species with excluded-volume effects. *The Journal of Chemical Physics*, 137(20):204116, 2012.

[13] Barend Ubbo Felderhof. Diffusion of interacting brownian particles. *Journal of Physics A: Mathematical and General*, 11(5):929, 1978.

[14] Hyun-Myung Chun, Xavier Durang, and Jae Dong Noh. Emergence of nonwhite noise in langevin dynamics with magnetic lorentz force. *Physical Review E*, 97(3):032117, 2018.

[15] Iman Abdoli, Hidde Derk Vuijk, Jens-Uwe Sommer, Joseph Michael Brader, and Abhinav Sharma. Nondiffusive fluxes in a brownian system with lorentz force. *Physical Review E*, 101(1):012120, 2020.

[16] Hidde Derk Vuijk, Jens-Uwe Sommer, Holger Merlitz, Joseph Michael Brader, and Abhinav Sharma. Lorentz forces induce inhomogeneity and flux in active systems. *Physical Review Research*, 2(1):013320, 2020.

[17] Iman Abdoli, Hidde Derk Vuijk, René Wittmann, Jens-Uwe Sommer, Jospeh Michael Brader, and Abhinav Sharma. Stationary state in brownian systems with lorentz force. *Physical Review Research*, 2(2):023381, 2020.

[18] Iman Abdoli, Erik Kalz, Hidde Derk Vuijk, René Wittmann, Jens-Uwe Sommer, Joseph Michael Brader, and Abhinav Sharma. Correlations in multithermostat brownian systems with lorentz force. *New journal of physics*, 22(9):093057, 2020.

[19] Jan Karel George Dhont. *An Introduction to Dynamics of Colloids*. Elsevier, 1996.

[20] Sarwat Hanna, Walter Hess, and Rudolf Klein. Self-diffusion of spherical brownian particles with hard-core interaction. *Physica A: Statistical Mechanics and its Applications*, 111(1–2), 181–199, 1982.

[21] Bruce J. Ackerson and L. Fleishman. Correlations for dilute hard core suspensions. *The Journal of Chemical Physics*, 76(5):2675–2679, 1982.

[22] The many-body Brownian-dynamics simulations were performed by Hidde D. Vuijk and Iman Abdoli at the Leibniz Institute of Polymer-Physics, Dresden in collaborative work for the article to be published on the findings of this work. I express my thankfullness to them for allowing me to use the genereted data in this thesis.

[23] Paul Langevin. Sur la théorie du mouvement brownien. *Compt. Rendus*, 146:530–533, 1908.

[24] Venkataraman Balakrishnan. *Elements of Nonequilibrium Statistical Mechanics*. Springer, 1 edition, 2008.

[25] Hannes Risken and Till Frank. *The Fokker-Planck Equation*, volume 18 of *Springer Series in Synergetics*. Springer-Verlag Berlin Heidelberg, 2 edition, 1996.

[26] Crispin Gardiner. *Stochastic Methods*, volume 13 of *Springer Series in Synergetics*. Springer-Verlag Berlin Heidelberg, 4 edition, 2009.

[27] Pavel Grinfeld. *Introduction to Tensor Analysis and the Calculus of Moving Surfaces*. Springer-Verlag, New York, 1 edition, 2013.

[28] David Gilbarg and Neil Sidney Trudinger. *Elliptic Partial Differential Equations of Second Order*, volume 224 of *Classics in Mathematics*. Springer-Verlag Berlin Heidelberg, 2 edition, 2001.

[29] William Henry Press, Brian Paul Flannery, Saul Arno Teukolsky, and William Thomas Vetterling. *Numerical Recipes*. Cambridge University Press, 3 edition, 2007.

[30] Sarwat Hanna, Walter Hess, and Rudolf Klein. The velocity autocorrelation function of an overdamped brownian system with hard-core intraction. *Journal of Physics A: Mathematical and General*, 14(12):L493, 1981.

[31] Thomas E Hull and Charlotte Froese. Asymptotic behaviour of the inverse of a laplace transform. *Canadian Journal of Mathematics*, 7:116–125, 1955.

[32] Alexander Juljewitsch Grosberg and Jean-François Joanny. Nonequilibrium statistical mechanics of mixtures of particles in contact with different thermostats. *Physical Review E*, 92(3):032118, 2015.

Printed in the United States
by Baker & Taylor Publisher Services